Kansei Innovation

Practical Design Applications for Product and Service Development

Industrial Innovation Series

Series Editor
Adedeji B. Badiru
Air Force Institute of Technology (AFIT) – Dayton, Ohio

PUBLISHED TITLES

Carbon Footprint Analysis: Concepts, Methods, Implementation, and Case Studies,
  Matthew John Franchetti & Defne Apul

Communication for Continuous Improvement Projects, *Tina Agustiady*

Computational Economic Analysis for Engineering and Industry, *Adedeji B. Badiru &
  Olufemi A. Omitaomu*

Conveyors: Applications, Selection, and Integration, *Patrick M. McGuire*

Culture and Trust in Technology-Driven Organizations, *Frances Alston*

Global Engineering: Design, Decision Making, and Communication, *Carlos Acosta, V. Jorge Leon,
  Charles Conrad, & Cesar O. Malave*

Handbook of Emergency Response: A Human Factors and Systems Engineering Approach,
  Adedeji B. Badiru & LeeAnn Racz

Handbook of Industrial Engineering Equations, Formulas, and Calculations, *Adedeji B. Badiru &
  Olufemi A. Omitaomu*

Handbook of Industrial and Systems Engineering, Second Edition *Adedeji B. Badiru*

Handbook of Military Industrial Engineering, *Adedeji B. Badiru & Marlin U. Thomas*

Industrial Control Systems: Mathematical and Statistical Models and Techniques,
  Adedeji B. Badiru, Oye Ibidapo-Obe, & Babatunde J. Ayeni

Industrial Project Management: Concepts, Tools, and Techniques, *Adedeji B. Badiru,
  Abidemi Badiru, & Adetokunboh Badiru*

Inventory Management: Non-Classical Views, *Mohamad Y. Jaber*

Kansei Engineering - 2-volume set
  * Innovations of Kansei Engineering, *Mitsuo Nagamachi & Anitawati Mohd Lokman*
  * Kansei/Affective Engineering, *Mitsuo Nagamachi*

Knowledge Discovery from Sensor Data, *Auroop R. Ganguly, João Gama, Olufemi A. Omitaomu,
  Mohamed Medhat Gaber, & Ranga Raju Vatsavai*

Learning Curves: Theory, Models, and Applications, *Mohamad Y. Jaber*

Managing Projects as Investments: Earned Value to Business Value, *Stephen A. Devaux*

Modern Construction: Lean Project Delivery and Integrated Practices, *Lincoln Harding Forbes &
  Syed M. Ahmed*

Moving from Project Management to Project Leadership: A Practical Guide to Leading Groups,
  R. Camper Bull

Project Management: Systems, Principles, and Applications, *Adedeji B. Badiru*

Project Management for the Oil and Gas Industry: A World System Approach, *Adedeji B. Badiru &
  Samuel O. Osisanya*

Quality Management in Construction Projects, *Abdul Razzak Rumane*

Quality Tools for Managing Construction Projects, *Abdul Razzak Rumane*

Social Responsibility: Failure Mode Effects and Analysis, *Holly Alison Duckworth &
  Rosemond Ann Moore*

Statistical Techniques for Project Control, *Adedeji B. Badiru & Tina Agustiady*

PUBLISHED TITLES

STEP Project Management: Guide for Science, Technology, and Engineering Projects,
   Adedeji B. Badiru

Sustainability: Utilizing Lean Six Sigma Techniques, *Tina Agustiady & Adedeji B. Badiru*

Systems Thinking: Coping with 21st Century Problems, *John Turner Boardman & Brian J. Sauser*

Techonomics: The Theory of Industrial Evolution, *H. Lee Martin*

Total Project Control: A Practitioner's Guide to Managing Projects as Investments,
   Second Edition, *Stephen A. Devaux*

Triple C Model of Project Management: Communication, Cooperation, Coordination,
   Adedeji B. Badiru

FORTHCOMING TITLES

3D Printing Handbook: Product Development for the Defense Industry, *Adedeji B. Badiru
   & Vhance V. Valencia*

Cellular Manufacturing: Mitigating Risk and Uncertainty, *John X. Wang*

Company Success in Manufacturing Organizations: A Holistic Systems Approach,
   Ana M. Ferreras & Lesia L. Crumpton-Young

Essentials of Engineering Leadership and Innovation, *Pamela McCauley-Bush &
   Lesia L. Crumpton-Young*

Global Manufacturing Technology Transfer: Africa-USA Strategies, Adaptations, and Management,
   Adedeji B. Badiru

Guide to Environment Safety and Health Management: Developing, Implementing, and
   Maintaining a Continuous Improvement Program, *Frances Alston & Emily J. Millikin*

Handbook of Construction Management: Scope, Schedule, and Cost Control,
   Abdul Razzak Rumane

Handbook of Measurements: Benchmarks for Systems Accuracy and Precision, *Adedeji B. Badiru
   & LeeAnn Racz*

Introduction to Industrial Engineering, Second Edition, *Avraham Shtub & Yuval Cohen*

Kansei Innovation: Practical Design Applications for Product and Service Development,
   Mitsuo Nagamachi & Anitawati Mohd Lokman

Project Management for Research: Tools and Techniques for Science and Technology,
   Adedeji B. Badiru, Vhance V. Valencia & Christina Rusnock

A Six Sigma Approach to Sustainability: Continual Improvement for Social Responsibility,
   Holly Allison Duckworth & Andrea Hoffmeier Zimmerman

Kansei Innovation

Practical Design Applications for Product and Service Development

Mitsuo Nagamachi
Anitawati Mohd Lokman

CRC Press
Taylor & Francis Group
Boca Raton   London   New York

CRC Press is an imprint of the
Taylor & Francis Group, an **informa** business

This book has been previously published in Japanese by Kaibundo Publishing (Tokyo).

CRC Press
Taylor & Francis Group
6000 Broken Sound Parkway NW, Suite 300
Boca Raton, FL 33487-2742

© 2015 by Taylor & Francis Group, LLC
CRC Press is an imprint of Taylor & Francis Group, an Informa business

No claim to original U.S. Government works

Printed on acid-free paper
Version Date: 20141124

International Standard Book Number-13: 978-1-4987-0682-7 (Paperback)

Library of Congress Cataloging-in-Publication Data

Nagamachi, Mitsuo, 1936-
    Kansei innovation : practical design applications for product and service development
/ authors, Mitsuo Nagamachi and Anitawati Mohd Lokman.
        pages cm -- (Industrial innovation series ; 32)
    Summary: "The idea of Kansei engineering was conceived in 1970. Since then, the
founder has worked to establish the Kansei engineering methodology and has created
more than fifty new product developments. Kansei engineering research is gaining
popularity in many countries"-- Provided by publisher.
    Includes bibliographical references and index.
    ISBN 978-1-4987-0682-7 (paperback)
    1. New products. 2. Human engineering. I. Lokman, Anitawati Mohd. II. Title.

TS170.N34 2015
658.5'75--dc23   2014032038

Visit the Taylor & Francis Web site at
http://www.taylorandfrancis.com

and the CRC Press Web site at
http://www.crcpress.com

Contents

Preface

A person who understands another person's intention and perfectly responds to it is normally regarded as a person who understands sensibility. If such a person is part of a service staff, customers will patronize the shop with satisfaction. In this case, we can say that the staff has proficiently sensed the customer's feelings and responded in a manner to satisfy the customer; in other words, the staff possesses *Kansei*. Kansei is the sensitivity that anybody can have. *Sensitivity* means the ability to understand what a customer wants by sensing it from the customer's eyes, facial expression, spoken words, etc. This sensitivity or Kansei can be acquired by anybody through training. And it is also possible to develop a new product that will be appreciated by customers, by studying, monitoring, and analyzing customers' Kansei. This method is called *Kansei engineering*.

The idea of Kansei engineering was conceived in 1970. Since then, Mitsuo Nagamachi, the founder, has worked to establish the methodology of Kansei engineering, and has assisted in developing no less than 50 new products. Today, Kansei engineering has spread to various parts of the world, and Kansei engineering research is gaining popularity in many countries. Kansei engineering is a scientific method where new products are developed, starting from analyzing customers' Kansei. Since new products are built based on customers' Kansei data, the new products will be well received by the customers, and thus become profitable products. We can also analyze what kind of merchandise displays are preferred by customers, or what kind of services are appreciated by customers, from the data of customers who have visited a shop. Kansei analysis can also be useful in developing service innovations.

Kansei-rich people can also be found in factories. They are the employees of the factories. What motivates them and how can we treat them better? In workplaces where efficiency has stalled, or where accidents occur, there are working methods that are inconsistent with the employees' Kansei, and there are also problems with the environment. These workplaces will improve if we consider the issues from Kansei and make some changes. These issues are also in the domain of ergonomics, but if we analyze them from both ergonomics and Kansei engineering perspectives, the processes will be much easier to work on, and workplaces will experience higher productivity. With industrial engineering (IE) improvement, we can realize a workplace improvement with a difference.

Think about the workers or employees as a whole. If we consider the Kansei of all the people at work, we might be able to even create an organizational form that differs from organization theory. Such an organization will exhibit increased motivation and faster decision making. By utilizing Kansei, new organizational developments will be possible. If we use this principle and methodology, city planning for a small community will also be possible. Perhaps it will become a brighter and better community to live in, where its residents are bound by strong emotional ties.

Nagamachi, the founder of Kansei engineering, first studied psychology when he obtained his Ph.D. in literature (theoretical psychology) in 1963. While in graduate school, he studied medical science and engineering. In medical science, he focused especially on cerebrophysiology and conducted pathological research into Alzheimer's disease. In 1967, he was invited to the University of Michigan Transportation Research Institute as an ergonomics researcher, and conducted joint research with GM, Ford, and Chrysler. There he studied the state-of-the-art automotive technology at that time. With this background, after returning to Japan, he was requested by the Ministry of International Trade and Industry to sit on the Automotive Ergonomics Research Committee as the person in charge of ergonomics, where he worked on raising Japan's automotive players, Toyota, Nissan Motor, Honda, Mitsubishi Motors, Mazda, and others, to a world-class level. Later, he played a role in mentoring countless Japanese corporations, such as Nippon Steel, Sumitomo Metal Industries, Mitsubishi Heavy Industries, Komatsu, Matsushita Electric Industrial, Matsushita Electric Works, Kubota, Daikin, Hitachi, Toshiba, etc., in the fields of industrial engineering and quality management.

In the 1970s, Nagamachi established Kansei engineering and helped many companies in new product development. Especially for Matsushita Electric Works (currently Panasonic Electric Works), he provided guidance in developing most of their household products, such as roof tiles, gutters, siding, bathtubs, kitchens, toilets, and so on. Kansei engineering has spread worldwide.

In this book, Nagamachi shares his 50 years of experience in enterprise guidance and product development using Kansei analysis, and includes examples of exceptional innovations of their time. You may already be aware of some of these innovations, and some might surprise you. You will understand how Kansei is analyzed scientifically and how it is applied in multilateral contexts.

The contents of this book are stories taken from real life. Among the companies mentioned are Nissan Motor, Mazda, Toyota, Volvo, Fuji Heavy Industries, Mitsubishi Electric, Tenmaya Department Stores, Seibu Department Stores, Suntory, NEC, Sharp, Komatsu, Wacoal Corporation, Matsushita Electric Works (Panasonic Electric Works), Boeing, and many more.

Finally, the author will explain how to nurture Kansei and develop the skills necessary for observing customer behavior. The goal of this book is to educate employees in Kansei.

Mitsuo Nagamachi
Founder of Kansei Engineering

About the Authors

Mitsuo Nagamachi, Ph.D., was born in Kobe in 1936 and received his Ph.D. (psychology) from Hiroshima University in 1963. His work experience includes professor of Faculty of Engineering at Hiroshima University, visiting ergonomist at the University of Michigan's Transportation Research Institute (TRI), president of Kure National Institute of Technology, dean, Kansei Design Department at Hiroshima International University, and visiting professor of User Science Research Institute at Kyushu University. He is currently serving as CEO of the International Kansei Design Institute, specializing in product development and quality control, safety engineering, and Kansei engineering.

Nagamachi founded Kansei engineering in the 1970s and engaged in new hit product development for a number of companies, such as Toyota, Nissan, Mazda, Isuzu, Komatsu, GM, Ford, Sharp, Wacoal, Mizuno, Milbon, Panasonic Electric Works, Nestle, P&G, and Johnson & Johnson.

In 1992, Nagamachi received the Distinguished Foreign Researcher Award from the Human Factors Society (United States) and, in 2002, the Fellowship Award from the International Society of Ergonomics. Nagamachi received the 2008 Minister of Education, Culture, Sports Award, "The Ministry of Science and Technology" (Division of the Advancement of Science). That same year, he was given the Chinese Culture Award, and, in 2012, the Emperor Prize, Gold Rays with Neck Ribbon Medal. Nagamachi has written 95 books, including *Kansei Engineering*, 198 academic papers, and 200 international conference papers.

Anitawati Mohd Lokman, Ph.D., is an associate professor at Universiti Teknologi MARA Malaysia, and a pioneer of Kansei engineering technology in Malaysia. She was entrusted with the knowledge by the founder of the technology, Professor Mitsuo Nagamachi, via collaborations in research, coauthoring books and research articles, as well as conducting workshops and tutorials locally and internationally. She established the Malaysia Kansei Engineering Society and was appointed its founding president. Lokman is known in academia internationally and has often consulted in the fields of Kansei engineering, affective design, human factors, interaction design, and website interface design. She recently established the Kansei/Affective Engineering (KAE) Research Interest Group, which gathers active researchers involved in the convergence of Kansei engineering in diverse fields such as ethics, racial integration, intelligent systems, robotics, and news media. Lokman has written 4 books, more than 60 academic papers, and more than 50 international conference papers in the field of Kansei engineering.

1

What Is Kansei?

1.1 Everybody Has Kansei

When we run across a beautiful woman, our feelings say, "What a pretty woman!" When we see a woman in appropriate, graceful (not flashy) makeup and elegantly dressed, we feel that she has been raised in a good environment and has a sense of elegance. We make inferences about the person's gentle character, privileged home environment, and so forth. This kind of feeling is called Kansei. Kansei is the feeling felt by the receiver of stimuli contained in the atmosphere of a situation. If the receiver is rich with emotions, a feeling that matches the stimuli will come out, but if the receiver is lacking in emotions or being defiant, he or she can only respond to a portion of the stimuli, and the feeling will be distorted. When a child who loves animals finds a puppy on a roadside, he or she will hug the puppy with sparkling eyes, but another child who has no such emotion will chase the puppy around with a stick. Even if the stimuli from the environment is the same, the emotion created by the stimuli, which is Kansei, may be different.

Different employees show different responses to customers, even though they receive the same new employee training and corporate education; some of them are very much liked by customers, while others very frequently cause trouble with customers. This difference is related to whether the person who received the education or training really understands the philosophy of the company. Therefore, there is a need for training that is easy to understand and has much practical work incorporated into it. In corporations that are recognized by the public as providers of excellent service, there is a training system that focuses on employees' emotions and values and nurtures their incorporation into the work process—this is Kansei education.

The functions that are the basis for Kansei are the five senses—eyesight, hearing, taste, touch, and smell—plus the deep sensation we get that is related to, for example, the ride quality of a car. Eyesight is the sense when something is visible to the eyes, and it is related to shape, color, tone, light, etc. These stimuli pass through the eyeball to touch the retina; the signals then reach the occipital lobe and pass through a few other regions before being processed as Kansei at the prefrontal cortex. Something that has a good shape will be interpreted as a beautiful or well-balanced feeling, and something that has colors and is well balanced will be sensed as an attractive thing. We do not sense it merely as a result of physiological processes, but this kind of Kansei is generated in combination with past experiences and memory (hippocampus) information. Therefore, Kansei that feels the beauty or attractiveness through eyesight alone will not be as meaningful to a person who does not have good experiences to incorporate in that feeling. The same thing for hearing; if we just sensuously feel the sound streaming from a CD, it will be merely a sound, but to a person who is familiar with classical music, he or she will surely be captivated by *Pastoral*, a symphony conducted by Furtwangler and composed by Beethoven. If the person's life philosophy and in-depth knowledge are incorporated in the physiological sensation, then a deep Kansei (savoring) related to symphony will come along.

Visual emotion is the external stimulation that passes through the eyes to the occipital lobe, interpreted as shape, tone, motion, etc., and the memories, knowledge, and experiences that were previously stored are added before the emotion reaches the prefrontal lobe. When a scene enters one's eyes, the person's accumulated visual database is added to the visual information; thus, he or she will see "delicious-looking food" or a "kindly person." The same thing applies to auditory stimulation. Sound stimulation enters the ears, which are sensory organs. While the stimulation is a mere physical sound wave, it is processed and flavored by the person's life experience, and by the time it reaches the prefrontal lobe, it will be heard as a "beautiful bird-song." All emotions through touch, taste, and smell are felt as the emotions supported by the person's daily life experiences. Composite emotions that combine sensations (sensory modalities), especially eyesight, which dominates 70%–80% of sensory stimulation, greatly impact Kansei; for example, eyesight + taste or eyesight + touch. The pain from being pinched by someone special will not be felt as a pain, and a photo of delicious-looking food is sufficient to make us sense how good it would taste. In addition to the five senses described above, there is the internal receptor, which gives us a composite sensation about something, for example, the ride quality of a car, but cannot be associated with a specific sensory organ.

Figure 1.1 shows how taste information received by taste buds inside our mouth (taste receptors) passes through a few pathways and is processed before the information that reaches the primary and secondary gustatory cortexes is interpreted into various tastes at the prefrontal cortex. As shown in the figure, sweetness, astringency, and other tastes pass through a complex pathway, intricately processed into taste sensation. At the same time, the experience information for the taste is added to the chemical information in the interpretation process. Emotions such as "Japanese plum tastes sour," "green tomato is not sweet," etc., will be generated and recorded into the long-term memory. Similarly, eyesight, hearing, smell, touch, etc., pass through a complex pathway before reaching the prefrontal cortex, together with respective emotions. Figure 1.2 shows the simplified neural pathways for eyesight and hearing.

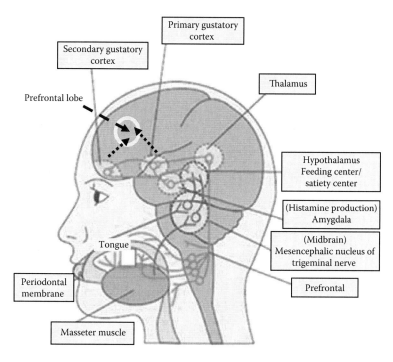

FIGURE 1.1
Pathway on which the taste sense is processed at the cerebral circuitry before arriving at the prefrontal cortex.

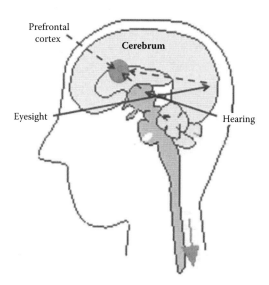

FIGURE 1.2
Information processing for eyesight and hearing is finally sensed as Kansei at the prefrontal cortex.

1.2 Let's Explore Kansei

Searching for what kind of product a customer wants from his facial expressions and motions, and judging whether a task is a heavy burden to a worker, or whether he or she is satisfied with his or her job based on his or her movements, are the Kansei issues dealt with in this book.

Mitsuo Nagamachi, the founder, has explained before that Kansei is composed of eyesight, hearing, smell, taste, touch, and an internal receptor. Respective sensory experiences are memorized together with the episode memory experienced at that time, and they are recorded as emotions. Something that is being memorized, decided, or judged as a certain value on top of these emotions by a function called *perception* will ooze out when a sensation similar to the previous ones is felt, and the prefrontal cortex will sense: "Ah, this one is very similar to the cooking that I have eaten before, so it will be delicious." The feeling of "how beautiful!" when a person saw the colored sunset described by Ken'ichiro Mogi in his book[1] is also a manifestation of a high-order function transmitted from eyesight (Figure 1.3).

The managerial staff of a company must include someone who understands the emotion of his or her subordinates toward the job by looking at their actions and expressions; a salesperson of a department store must have the ability to understand the emotion of visiting customers toward the lined-up merchandise by looking at their actions, facial expressions, and how they speak; and it is desirable for product development staff to accurately grasp social demands—what kinds of products are currently sought—from

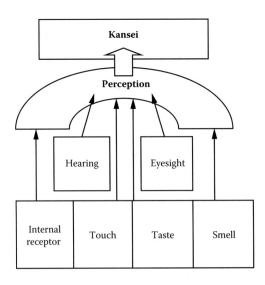

FIGURE 1.3
Kansei expressed after information from the six senses passes through the perception function.

customer information and develop an appreciated product. It is desirable for a schoolteacher to sense the symptoms of bullying early in the relationships between students and provide proper guidance. All these roles require figuring out concealed emotions and Kansei beneath the actions, expressions, spoken language, etc., of the persons they are dealing with. In order to do this, the managerial staff, customer service personnel, teacher, and so on, need to have rich and deep Kansei themselves, and have the ability to read Kansei of the persons they are dealing with.

1.3 Kansei Is Expressed in Various Forms and Shapes

Figure 1.4 shows how Kansei is represented in various forms and shapes. A person's emotional state can be measured physiologically. For example, if it is measured using brain waves or functional magnetic resonance imaging (fMRI), we can know the activity state of the cerebrum and the emotional state. We can check whether someone is performing an improperly tough job by measuring the state of the muscle using an electromyograph. A customer will definitely turn his eyes to the products that catch his interest, so we can measure his or her eyeball movement with an eye camera. We can know whether a food tastes good or bad by looking at the facial expression of the person who is eating the food. The difficulty of a job is proportional to the degree of hip bending and distance walked, and a worker's motivation can be sensed based on the energy that the worker exhibits. And when someone

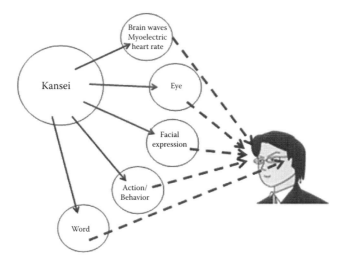

FIGURE 1.4
Various forms of Kansei expression.

is deeply moved, he or she will express it using words. He or she will utter something like "That's cool!" "It was so delicious!" or "That's uncool!" so we can know his or her Kansei conditions. By being alert to these kinds of expressions, we will be able to apply human Kansei to product development or work improvement.[2]

There are ergonomics and Kansei engineering methods to study human Kansei that should not be overlooked. In brief, they are as follows:

1. As a method for product development:
   a. Observational method. Observe customer actions for a while to search for certain characteristics in body and eye movements. As necessary, we may also need to count these movements. If a person looks at, sniffs, touches, or returns to a product a few times, we will know that he or she is attracted to something about that product.
   b. Kansei analysis. After we have grasped the overall interest level, we should conduct a Kansei survey and analysis of what captured his or her interest—is it shape, color, taste, or price? Depending on the first Kansei prehension, prepare a list of 40–60 important Kansei words related to the survey and conduct a survey of customers. Put the obtained data through a statistical analysis (multivariable analysis) to understand the factors that make up the Kansei. Use the results of the statistical analyses from a few different product Kansei surveys to develop a new product.
   c. Cerebrophysiological method. Nagamachi earlier described that Kansei is the emotion generated from the prefrontal cortex. In order to know what is happening inside the cerebrum, we need the cerebrophysiological method. Methods that are commonly used are fMRI, NIRS, EEG, etc.
   d. Artificial intelligence. In Kansei engineering, the data acquired from the Kansei analysis is used as an artificial intelligence rule base and inference engine. When we set the design that we are looking for as the objective function, and enter the appropriate Kansei words, an ideal design will be generated. There are a wide variety of artificial intelligence methods.
   e. Design from marketing. There is also a method where the required data for a new product are extracted by analyzing the data acquired from surveys, before commercializing the product. The use of *text mining*, and recently *big data*, are popular.
2. As a method of job improvement:
   a. Industrial engineering improvement. In an effort to maintain production efficiency and quality, we seek to improve the work flow or consider the work distribution between robots and humans.

Such systems include modules where parts are preassembled into units, as an effort to reduce assembling time. Among them, the Toyota Production System is an excellent example.

b. Ergonomics improvement. This refers to improvements that make work easier by considering physical limitations, reducing action energy, or sufficiently incorporating human strengths and weaknesses.

c. Kansei ergonomics improvement. This method motivates workers, thus bringing pleasure and a sense of purpose in work, and as a result, production efficiency will improve. The cell system is one such example.

d. Service innovation. This is a method that not only tries to introduce the concepts and techniques of an engineering system in the service industry, but also focuses on customer Kansei and develops advanced services that maximize understanding of customer Kansei and customer satisfaction.

References

1. K. Mogi. *Brain and qualia* (in Japanese). Tokyo: Nikkei Science, 1997.
2. M. Nagamachi. *The story of Kansei engineering* (in Japanese). Tokyo: Japanese Standards Association, 1995.

2

Applying Kansei to Manufacturing Technology

2.1 The Birth of the Doorless Assembly Line

When Mitsuo Nagamachi started working with the Aging of Society Policy, by then the Labor Ministry Employment Security Bureau from the 52nd year of the Showa era (1972), he had been mentoring Toyota Motor and Nissan Motor on aged-persons employment strategy. During his lecture on the Aging of Society Policy to 70 shop floor industrial engineering (IE) staff at the Nissan Motor Zama plant, as an example, he gave advice about improvement of the final fitting line. At the time, the fitting line was only for workers under 40 years of age. If many aged workers entered the workforce, there would be a higher possibility that an aged worker would be assigned to the fitting line, requiring them to do assembly work in unusual and awkward positions. The discovery of physically difficult work in the workplace is the duty of *ergonomists*.[1,2]

He felt that he could use the results of experiments related to energy consumption of the working posture relative metabolic rate (RMR), which he had been conducting in Hiroshima University's ergonomics lab under a research grant from the Labor Ministry.[3,4] He had thoroughly looked into what kind of working postures there were at all production lines at Nissan Motor, and compiled them into 10 kinds of typical postures in the automotive industry. Then he asked his students to assume and hold those postures, and he conducted a job aptitude test—"bean picking"—for 3 minutes. During the test, the students were made to carry a Douglas bag on their backs while an electromyogram was taken of their erector muscles. It was very heavy labor for the students, but as Nagamachi elaborated later, this study resulted in very valuable data.

The left side of Figure 2.1 shows 10 kinds of typical working postures that were compiled at that time. Working postures include sitting and standing positions, going into a bending posture as bending at the waist and kness increases, and finally, where bending is greatest, like squatting on a toilet.

Class.	Score	Posture	Descriptions	Specific example
J	10		Half-crouching with knees deeply bent, and upper body bent forward	Heels are floating (Like swimming race start)
I	6		Half-crouching with straight knees, and upper body deeply bent forward	≥90° Also same if knees bent in this posture
H			Half-crouching with knees bent, and upper body bent forward	45~90° 0~45°
G			Half-crouching with straight knees, and upper body bent forward	45~90° Also same if there is an obstruction at one's legs
F	5		Squatting posture (Heels on the ground)	If heels float, knees will stick out forward ---Classification (J)
E			Upper body slightly bent forward with knees straight	30~45° If look as strained posture, Classification (G)
D	4		Upper body slightly bent forward with knees slightly bent	0~30° Knees slightly bent in standing posture
C	3		Overreaching in standing posture	Like reaching out for something above eye level
B	1		Standing posture	0~30° Back muscle is straight
A			Sitting posture	Including kneeling posture

FIGURE 2.1
Classification of working postures.

The postures in the lower part of the diagram have a small energy metabolic rate and generate less myoelectricity, and these values increase as we go higher in the diagram.

The result in graph form is shown in Figure 2.2. The greater the bending of the waist and knees, the larger the RMR value, indicating tough work. A general rule for the difficulty in working posture is that there's almost no difference between sitting and standing postures, but a gradual bending of the knees will increase the difficulty, as will an increase in waist bending. When the waist and knees are bent at a same time, the posture becomes more difficult.

Let's get back to what had happened at Nissan Motor. Nagamachi spoke to the 70 IE staff: "When the car body enters the fitting line after painting is finished, remove all four doors at the first process, and flow them into the dedicated door assembly line, where only the doors will be assembled using highly efficient devices. If the interior fitting on the empty body with the doors removed is performed in a comfortable posture, even the aged

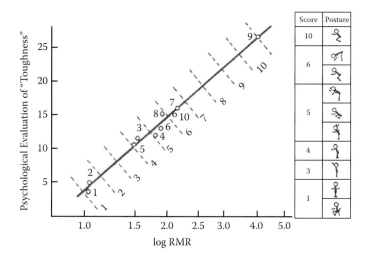

FIGURE 2.2
Relation between RMR of each posture and toughness.

workers can do it. We return the removed doors, automatically if possible, to their original body at the final process using a computer control, right?" All the IE staff had opposed this idea. The reason was that the removal and then return of the already fitted doors would create two new processes, which would reduce the work efficiency. That was a legitimate concern from the IE staff's perspective.

Nagamachi therefore gave a clear explanation of the energy efficiency of the new working postures: The distance between the parts box and car body would become shorter if no door were attached, the overall walking distance would be reduced, and work efficiency would improve due to the removal of the doors. In addition, at the dedicated door line, the door assembly work would dramatically improve by modifying the jig for the hanging door to a universal jig that could freely turn from a vertical to a horizontal position, and eventually there would be potential to improve working hours by 30–40%. He later provided them with drawings for the line change and noted on them the estimated efficiency improvements involved with the change.

A clear-sighted Fujii section head (the person who was eventually promoted to vice president) informed Nagamachi, "Your proposal makes sense. We will consider whether or not to change the line after conducting experiments." They put a car body in an empty space in the plant and conducted the experiment per his proposal. The results were exactly as he had said, so in 1977, the doorless fitting line was started in the Zama plant. When he talked about this at Mazda, the plant manager at that time, Mr. Kobayashi, took a great deal of interest in it. With study tours to Nissan Motor and Nagamachi's coaching, a doorless line was introduced at Mazda

1 year later. He provided guidance on the introduction of the doorless line during the establishment of the Toyota Tahara plant, and Honda also changed to the new fitting line.

Figures 2.3 to 2.5 are the idea drawings proposed to Nissan Motor. Figure 2.3 shows where the four doors are removed from the painted car body and sent to a dedicated door line. Figure 2.4 shows where the doors are assembled in an easy-to-fit manner using universal jigs at the dedicated door line. Figure 2.5 shows where the computer-controlled doors are fitted to their original car body in the final line process.[3]

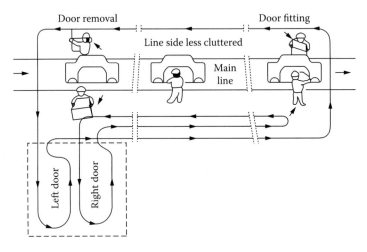

FIGURE 2.3
Removing and sending doors to another door line.

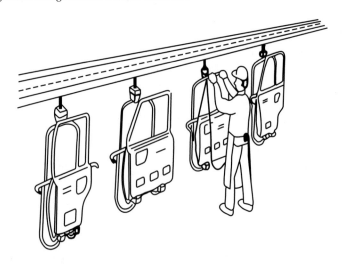

FIGURE 2.4
Dedicated door assembly line.

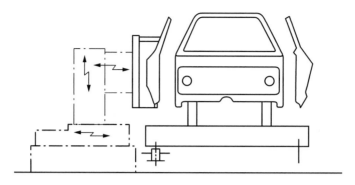

FIGURE 2.5
Automatic door installation system at the final process.

A few years after the introduction of the doorless fitting line (doorless assembly system (DAS)), which was also called the Nagamachi System at that time at Nissan Motor,[5] Nagamachi had a chance to conduct a simultaneous comparative study between the old line, which had doors attached, and the new doorless new line. The study revealed many outcomes, such as:

1. Due to the removal of four doors, the part rack could be brought infinitely closer to the car body, thus shortening the walking distance to pick up parts.
2. Due to no doors, interior fitting work became much easier.
3. Loading of the dashboard, seat, and other components using a robot became easy.
4. A trolley could be installed along the sides of the assembly line, so that workers could work on it, which reduced the walking distance per worker to less than 50%.

With those great benefits, upon discussions between employees and employer, it was decided to decrease the workstaff by 24 workers (which is the equivalent of 12%). As a result, the amount invested in the new line could be amortized in 10 months (Table 2.1).

From Figure 2.6, we can see that in the case of DAS, the position of the parts rack is considerably closer to the car body than it was in the old line. Figure 2.7 shows the improvement of working postures at the dedicated door line. The vertical axis represents the cumulative posture percentage, while the left side of the lower horizontal axis shows a severe bending posture; the severity becomes less toward the right side. Lastly, the figure shows a standing posture. There are two lines shown in the figure: the dotted line shows the cumulative working posture curve for the old system, while the solid line shows the cumulative posture curve for the new system. The sudden rise of the dotted line at the left side, followed by a gradual

TABLE 2.1

Effects of Doorless Fitting Line

No.	Item	Effect
1	Distance to part racks	Reduction of 0.66 m × 4 = 2.64 m/head
2	Interior installation	Became very easy
3	Loading of dashboard, seats, etc.	Robotization
4	Traveling jigs and certain parts	Extreme reduction of walking distance
5	Walking distance per worker	Reduction of more than 50%
7	Fitting of doors at a separate line	Total elimination of bending posture
8	Labor savings due to efficiency increase	Labor savings of 24/200 workers (12%)
9	New line invested cost (about JPY 200 million)	Amortized in 10 months

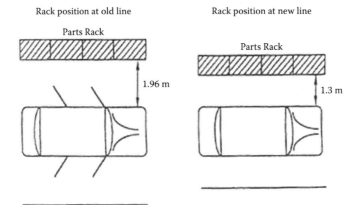

FIGURE 2.6

Comparison of part rack position at old and new fitting lines.

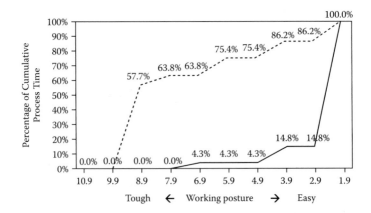

FIGURE 2.7

Old line door assembly (dotted line) has many bending working postures, while almost all working postures at the new line (solid line) are standing postures.

FIGURE 2.8
Volkswagen's doorless line.

increase toward 100%, indicates frequent occurrences of severe bending working postures; on the other hand, the solid line shows that most of the working postures are standing.

A long time ago, Toyota introduced a revolutionary change, where a chair was mounted to the tip of a large robot arm, a worker climbed onto it, and the arm took him into the car body. He performed the interior installation while riding the robot arm. After the operation was complete, the arm took the worker out. This was made possible thanks to the removal of doors and a wide car body.

Overseas, the system was transferred from Mazda to Ford, followed by GM and Chrysler; the transfer to Volvo was also quick. Due to the aforementioned benefits, now there isn't a single automotive maker that has not introduced this system. In fact, it is perceived as common sense. Upon seeing workplaces in which workers are having tough times, if we carry out improvement based on the understanding on their emotions (Kansei), we will get good results, thus making it spread worldwide. Figure 2.8 shows the doorless line in a Volkswagen.

2.2 Human-Centered Production Line

2.2.1 What Is Humanity?

Nagamachi thinks those who have watched *Modern Times*, a Charlie Chaplin movie, are able to understand how the simple flow operation erodes human hearts. Compared to the Ford production system of automotive assembly, the current automotive industry has been modernized considerably, but in any era, the relation between the production system and humanity has been pointed out by many researchers.

The first man–machine theoretical antithesis of Taylorism that attracted lots of attention is found in the Hawthorne studies, which were conducted by Roethlisberger and Dickson for 8 years, starting in 1924. Elton Mayo of Harvard, who led the studies, concluded that for workers, humanity (Kansei) is sentiment (emotion), and the characteristic that sustains a workplace is "human relations."

Since the basic idea of the study was that physical conditions have an effect on work efficiency, the conditions were changed by adjusting break times from 5 minutes to 15 minutes, reducing finish time from 60 minutes to 30 minutes, or setting the workplace lighting to extremely dark or bright, according to a psychological experiment plan. Despite those changes, work efficiency increased independent of the conditions. Professor Mayo, upon seeing this result, concluded that the result was due to the excited feeling (sentiment) of the factory girls participating in the experiment, as well as the cooperativeness and very good human relations between the supervisor (an admired Harvard undergraduate) and the workers.[6]

Another eminent study related to humanity is on coal mining and loading work, by E.L. Trist's group of Britain's Tavistock Institute. Skilled workers in coal mining, loading, conveying, etc., at open-pit coal mines were extremely cooperative with their master, and even in real life, they lived in a group around the master and his wife. However, when the coal mining method was mechanized, changing conventional burden sharing to a specialized style of work, troubles started to occur frequently. Trist's group concluded that the mechanized process had destroyed the past cooperative human relations. Trist's group positioned this as the first study on a "social and technical system (sociotechnical system)," and said that the social system (human system) and the technical system (production system and mechanization) need to be designed in a manner where they are successfully interfaced with each other.

Next, is the well-known human needs system of A.H. Maslow. After conducting many studies related to human needs, Maslow pointed out that humans build up five needs in phases, and promulgated the theory of a five-level hierarchy of needs (Figure 2.9). Higher needs are not satisfied until lower-level needs are satisfied. For humans, when the low-level needs are satisfied, the esteem needs become more important, stimulate personal growth, and motivate personal development to reach the highest level, self-actualization. The willingness to take up the challenges toward one's personal objective with a sense of responsibility will appear. The theory says that progressing to the higher needs is not solely the responsibility of the worker, but requires considerations of the workplace atmosphere, personnel management, the contents of the job, etc.[4]

There is a famous theory by Frederick Herzberg, the motivator-hygiene two-factor theory. He conducted field surveys on many groups, such as

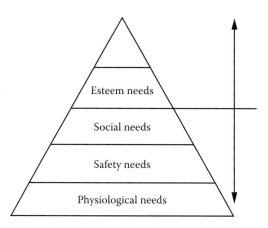

FIGURE 2.9
Maslow's five-level hierarchy of needs.

accountants, engineers, etc., and presented dazzling results related to work conditions when motivation was produced (satisfaction) and when motivation was not produced (dissatisfaction). He explained that workers are motivated by achievement, recognition, the job itself, responsibility, etc., as per the right side of Figure 2.10; the factor not related to motivation (hygiene factor) is related to company policy, supervision method, relation with superiors, relation with colleagues, working conditions, etc.

A range of similar surveys was also conducted in Japan, and the findings were basically the same. However, for Japanese, human relations with superiors and colleagues are part of the motivator. Also, the important point is the job itself. In a short-segmented flow operation, neither responsibility nor sense of achievement will develop. In other words, providing a job that gives one a chance to prove himself or herself and a sense of responsibility on the job, is a condition for bringing out motivation.

The last theory that be must mentioned here is the design of job theory by Professor Louis E. Davis from the University of California. Through studies at Tavistock Institute, etc., Davis stressed that the harmonization of the technical system and the human/organizational system is necessary to maintain humanity in the workplace—the scenes where human work is full of machines and technology. As mentioned on the right, humans possess various humanities and, at the same time, do not really change or progress much. On the other hand, machines and technology keep on changing and progressing in line with the times. Therefore, the gap between the two just keeps on widening. When the progress of technology does not match that of humanity, maladaptation will occur on the human side. Consideration of how to preserve humanity, no matter how technology progresses—in other words, designing the work aspect to match humanity—is the meaning of job design (design of job). A creative design is necessary. The human

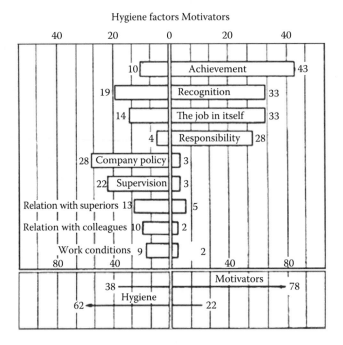

FIGURE 2.10
Herzberg's motivator-hygiene two-factor theory.

aspect is developed through the job, responsibilities are attached to jobs, and workers feel a sense of achievement when they fulfill them. A system where workers can evaluate themselves and their workplace is made available. The job itself is not monotonous, but complicated, with room for creativity.[5,6]

One such style is the cell production system, which is explained next.

2.3 Cell Production System

2.3.1 Cell Production Brings about the Purpose of Life and Occupational Satisfaction

In the Management Engineering Department of Hiroshima University's Faculty of Engineering, there is a conveyor belt for experimenting with operation flow. Nagamachi used it to conduct the following experiment. The objective of the experiment was to ergonomically analyze which production system gave the highest motivation—yielded a high productivity—among the conveyor operation, group operation, and one-man operation. Conveyor operation refers to an assembly flow operation with eight members; group operation is the flow operation by a pair of members (two pairs were used); and one-man operation refers to each of the eight members performing the whole assembly operation alone.

Nagamachi and his team asked a telephone set assembly company to let them borrow a large quantity of telephone set components, and the company agreed to buy back the products assembled by the students; the money was paid to the students as their hourly allowance. Twenty-four students received 1-week training at the company and acquired a skill level similar to that of the employees. The experiment started by adopting the same work regulations as the normal company, where work started at 8:30 a.m., with a 60-minute lunch break, and finished at 5:00 p.m. The experiment continued for 2 weeks. The data at the end of the first week were as shown in Figure 2.11. On average, the group with the highest productivity was the one-man operation, followed by the two-man operation, while the flow operation had the lowest productivity. In other words, the productivity of the conveyor belt operation was controlled by the belt speed (pitch time), so if the belt moved faster, the productivity increased, but worker satisfaction decreased. For the two-man operation, since they are coordinating with each other—even though they work while chitchatting—they become motivated. Since the one-man operation depends totally on competency and motivation, the person thinks that he or she is fully responsible for the completion of one telephone set unit; therefore, this brings out the highest performance. What is more, there is no necessity to apply pressure. At the end of the second week, there was no big change in the conveyor belt operation and the two-man group operation, but the one-man operation had increased up to the broken line in the following figure, and the

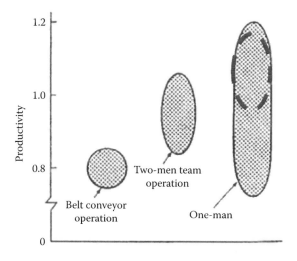

FIGURE 2.11
Productivity difference for the three modes of operation.

differences between individuals had also become smaller. This result is the effect provided by the mode of operation, which had considered humanity and the harmonization between work and humanity that Nagamachi had been explaining.[9]

Table 2.2 shows the result of a questionnaire survey conducted on the students regarding the feelings of job motivation and occupational satisfaction that they felt during this experiment. The "P" values indicate how the students felt about the work itself and their emotions. The numbers in the table show the ranking of the respective factors that influenced the external criteria. The most significant factor in job motivation is clarification, which refers to clearly seeing the scope of one's job responsibilities, which is greatest in the one-man operation.

Variation in job content is also a characteristic of the one-man operation. A negative effect on job motivation was skill level; this refers to the small work scope at the conveyor belt. Similarly, clarity with respect to the job responsibilities had a strong effect on satisfaction, which supports Herzberg's findings.[9]

This experiment shows that performing an entire unit task with greater responsibility brings about job motivation and satisfaction.

2.3.2 Fuji Heavy Industries One-Man Assembly Work

One day in 1974, a strange problem occurred in the assembly shop of Subaru Fourth Section at Fuji Heavy Industries, under the Nakano section manager. The younger group among 80 workers became disgusted with their jobs. The Nakano section manager could feel that the problem was not a mild one, so after work finished, he divided the workers into groups of six members and interviewed all of them. The workers voiced many quite serious anxieties, such as "can't get motivated because we're only tightening tire bolts," "we love cars, and in the future we want to open a car maintenance workshop, but our skills have not improved even though we have waited for a long time," etc. Therefore, the section manager, upon discussion with all the workers, came up with a proposal. "Now I know how you feel and what you want to do. I will get approval from the company president to organize an automotive technical class after work. Will all of you participate?" All of the workers agreed. Therefore, the Nakano section manager explained the situation to the company president, and with consent from the president, it was decided that the automotive technical class would be organized.

Then, in order to turn an uninteresting job into an interesting and satisfying job, he held a discussion with the workers, including the foremen. During the discussion, they proposed, "If each person assembles one complete car, we can understand all the skills, and that might be good for us too." Even though there was a bit of apprehension, since everyone agreed, he decided they would somehow conduct trials.

TABLE 2.2

Perception Study by Each Mode of Operation

External Criteria	P Change	P Mutual Connection	P Volunteerism Freedom	P Clarity	P Responsibility	P Skill Level	P Decision Making	P Visual Clarity	P into Processing Amount	Contributing Rate
					Factor					
Job motivation	7.122 (2)	2.659 (7)	4.216 (4)	9.503 (1)	1.984 (8)	1.714 (9)	2.823 (6)	5.508 (3)	3.001 (5)	0.966*
Job satisfaction	1.471 (5)	0.306 (9)	1.118 (6)	2.722 (1)	0.757 (7)	0.585 (8)	2.280 (2)	2.003 (3)	1.580 (4)	0.943*
Attendance rate	4.025 (3)	4.118 (2)	2.427 (5)	4.565 (1)	1.841 (6)	1.675 (8)	1.820 (7)	4.005 (4)	1.257 (9)	0.785*

Note: P denotes perception. Parentheses indicate ranking. Asterisks indicate a significant value above 0.5%.

To start with, he set up a group of six veteran workers from the straw boss class, with a plan for one worker to assemble one complete car; workers at each process would support the team, so he experimented to see how long it would take to achieve the one-man operation skill. The result was that 1 week was good enough. Since there was agreement that the result was not surprising because the workers were veterans, the next step was to select younger workers with poor performance; the group was called the "lucky seven"—to connote that they were "lucky to be selected"—and an experiment was conducted to see how long it would take for them to acquire the same level of skill as the straw bosses. At this time, with the workers at each station delightedly providing encouragement, and thanks to the skills coaching, surprisingly, the skill was acquired in just 10 days. For the other 68 workers, even with some trepidation, the training duration was approximated to be like that of these two groups. It took a few months to complete the one-man system training for all workers before they could fully embark on the one-man system—one worker, one car (Figure 2.12). Subsequently, the workers were divided into small groups of six, and each of the small groups was put on the assembly line and continued the one-worker, one-car production style. The six members who had completed the work on the line gathered in a room to discuss the locations of tools and components in the line where they had assembled the car, or the process improvement. These discussions helped in producing performance that was unparalleled in the world. Productivity increased 200%, the quality improved 100%, and cars were shipped out without requiring checking after they were "lined off."

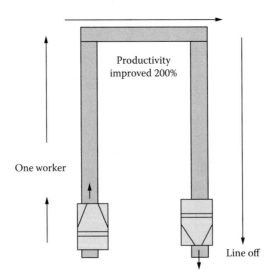

FIGURE 2.12
Subaru one-man system. One worker assembles one complete car.

2.3.3 Mitsubishi Electric JEL Production System

There are two types of job design: job enlargement, where small segments of short takt-time tasks are consolidated into a unit, and job enrichment, which makes the working hour for each person more varied and the unit bigger in anticipation of improvement in responsibility and skills. We can also rephrase job enlargement as a system where a process is spread horizontally, while job enrichment is a method of giving vertical depth.[7,8]

Around 1972 at Mitsubishi Electric's Fukuyama Works, grievances about work being monotonous and unmotivated were voiced by the workers at the non-fuse breaker assembly line. Process staff weighed those concerns heavily and discussed with the workers how to change the processes to improve the situation. As a result, they decided to change the present flow operation of nine workers to the tasks consolidation method, which has a smaller number of workers. In other words, it was a change to the job enlargement method. In the past, the takt time for each of the nine workers ranged from 40 seconds to 62 seconds and was not balanced. Therefore, as shown in Figure 2.13, it was lengthened to 4.5 minutes (270 seconds) for each worker. The new first process serves as the preceding process for the next process, which has four members, so it becomes a process of 4 minutes plus, while in the new third process, each of the members continues a portion of the task done by two members of the preceding process, so it will also exceed 4 minutes. These changes improved the line balance (this is called the JEL System).

Despite using two fewer workers than the old line, the productivity for the JEL line a year later had increased by 156%; the effect of the job consolidation system was great (Figure 2.14). Quality mistakes had gone down to a level of almost close to zero. Workers' opinions had also completely changed from the past, and their motivation and job satisfaction had improved.

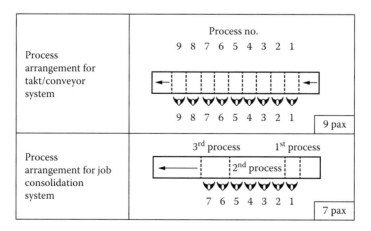

FIGURE 2.13
Old line (top) and JEL line (bottom).

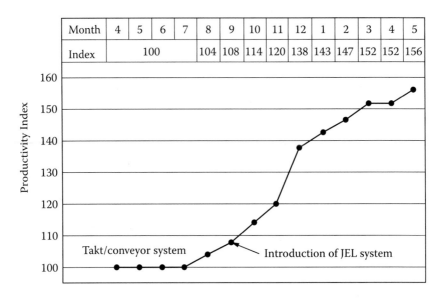

Month	4	5	6	7	8	9	10	11	12	1	2	3	4	5
Index	100				104	108	114	120	138	143	147	152	152	156

FIGURE 2.14
Productivity improvement based on job enlargement.

Since the Labor Union Morale Committee of Mitsubishi Electric was also in the midst of discussion about measures to further improve employees' job satisfaction, it was deeply interested in the success of Fukuyama Works, and gave instructions to all works to consider the matter.

The introduction of the JEL System was immediately considered at Nakatsugawa Works, where Mr. Abeyama, the chairman of the Morale Committee (graduate of the Faculty of Engineering, Hiroshima University), held the position of Production Engineering Department manager, and Nagamachi had been requested to cooperate as a researcher. Nakatsugawa Works was manufacturing big- and small-sized blower-related products, and it had been decided to apply a very unique "clean heater" manufacturing method. This product is a clean heater that applies the FF system using gas (the combustion gas is sent to outside the room), but due to the gas specifications, extra manpower was required because the gas detection personnel have to be stationed at each segment of the product assembly performed by 20 workers. After in-depth consideration, they decided to adopt the JEL System, where one worker would assemble the ignition part only, and the worker would conduct gas detection at every pivotal point of the assembly, changing the style to the so-called one-man system. Nagamachi conducted studies on the locations of parts and tools, and also conducted assembly training to improve the placement of materials in the system.

Figure 2.15 shows the old flow operation system in the upper row and the JEL Production System in the lower row. Twenty workers had lined up in

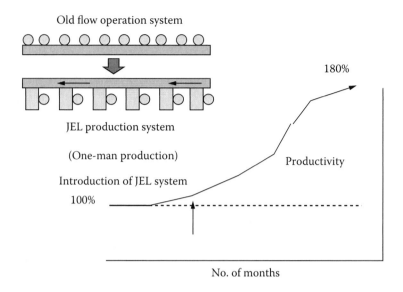

FIGURE 2.15
Productivity improvement with the introduction of the JEL Production System.

a row for the assembly process, but Nagamachi had 10 workers, each of them performing a complete assembly, line up in parallel along the flow operation. The productivity increased by a stunning 180% in 6 months. Of course, the satisfaction also improved. In the era of the Hayakawa plant manager, all workpieces were changed to the cell production system. In the succeeding era of the Abeyama plant manager, the cell production system was introduced in the die shop, which was totally unique indeed.

The die-making shop receives a design drawing from the design department, as well as explanations from the design staff. Cutting machines have respective machinists assigned to them, and it is like a flow process where machinists take turns according to the machining sequence until the die is completed. The production model is a flow operation. In this style, each machinist is only responsible for a small segment, so he or she will not be able to feel a sense of completion. As improvement measures, all die-making machinists underwent training in the design department for about a month, and intensive training was conducted until every machinist was fully able to operate all the machines. Once these preparations were completed, the cell production system for dies began. In the new system, the machinists have a briefing on the new order in the design department and prepare a simple design memo. They return to the die shop and prepare a machining process plan. One machinist completes all the machining processes by himself or herself using the necessary machines and tools. The average lead time for the conventional machine-specialized method was 36 days, but with the new system, the average lead time was reduced to one-sixth, i.e., 6 days. Of course, all members took pride in the completed dies.

At Mitsubishi Electric, the cell production system has been extended to Kamakura Works (lighting equipment), Shizuoka Works (refrigerators), and other plants. In the computer production industry, major companies have adopted the cell production system. Toshiba, NEC, Fuji Electric, etc., are known for this, and the system has been adopted in all Panasonic plants. At Daikin, even medium-sized air conditioners are assembled by one operator. And at Canon, it has been introduced in almost all plants; one worker assembles a larged-sized printer alone, and it is reported that the productivity has increased fivefold (Figure 2.16).[11] In the automotive field, which had been thought to be difficult, Volvo and Volkswagen are producing good results with the cell system (Figure 2.17).[9]

Traditional assembly line

Flower cell system

FIGURE 2.16
Canon changed to cell production.

Volvo two-men system

FIGURE 2.17
Volvo Devla plant's two-man system.

2.4 System That Creates Job Satisfaction and Purpose in Life

Professor Nagamachi thinks employees would like to happily and comfortably work. He cannot always say it is good work with a conveyor belt, but as mentioned in the previous section, there is responsibility in the work, a possibility of self-growth, a change, a chance of advancement, an element of accomplishment, and an aspect of being evaluated by oneself and others that are rewarding. Although conveyor belt work is not entirely bad, the simple task cycle times of 10–15 seconds may result in an unhealthy state of mind for workers who lack the job satisfaction that is part of the cell system.

Speaking from the spirit of the quality of working life (QWL) (increasing the quality of labor), because it is an important element in the life and work of the people, having the time is also significant. Therefore, to be embedded in the work of several elements, the elements listed above are desirable. The cell production system is one way to solve this problem.

In the cell production system, there are two different methods: job enrichment (mechanism to expand the business to the next level) and job expansion (mechanism that delves into the business and in particular the vertical side). The cell production system can be classified into two forms: one-person work (one-man system) and small-team work.

The important thing is, when considering completion of a product, if the process of completing it is not too complex, or the number of components is small, one-person work is possible.

In the case where the task is very complex and large numbers of components are involved, if we can break them down when we look at the whole process, it is a good idea to distribute the parts to workers in each unit of a dividing point. What must be considered here is that it must be a division into coherently meaningful units, and this will not happen by simply dividing units into tasks that are roughly equal in work time.

In operations such as assembling large automobile parts, rather than thinking it's impossible when it involves about 30,000 components, all manufacturers should complete most of the assembly work at the front of the line. This does not mean that all 30,000 components will be directly assembled at the end of the line.

A one-man operation is now possible in Fuji Heavy Industries Subaru line because the assembly is done at 63 large and small points. Since doorless module organization has been greatly progressing in the automotive industry recently, Volkswagen and Volvo have adopted one-man work or a two-person work system. Canon is similar.

However, although the format is a cell production system, and you often see a factory that supplies parts tactically, time constraints exist due to the takt conveyor supply, and thus it is not a true representation of the cell production system. As can be seen in the photos from Volvo, the part the worker brings to the cell production site is stacked. Depending on the work procedure of the operator, he or she carries his or her own parts from storage when the time separator is raised. This causes degrees of freedom for self-regulation of production speed to occur.

The second factor to enhance job satisfaction is to eliminate the "impossible" from work. The importance of working posture was mentioned earlier. It was mentioned that metabolic energy increases when the bending curve of the hip and knee increases, and myoelectricity of the spinae and thigh muscles increases. Work will be easier when we reduce bending at the knee and waist to near zero. This will improve the work that involves a standing posture. This study was conducted with research grants from the former Department of Labor, and thus the research results were released to the department. Ten types of working postures were stored in the cloud, and anyone can use the application and register, bring a mobile phone into the factory, have a look at one cycle of the work, check each time the working posture changes, one after another, and calculate the difficulty of posture (distress index). The work posture distress index is shown in Figure 2.18. If a work posture is more than 5 points, it must be improved so that it has 5 points or less.[10]

Let's look at an example that has achieved results. In 1982, Daikin Corporation, a specialized cooler maker, established a "friendly workplace" with a slogan "anyone can work from anywhere" as a job redesign program, to address the growth the number of aging of employees.

On the basis of consent from President Yamada, a joint research grant was received from the Employment of Older Persons Development Association (Foundation). Yamada Kanaoka factory's vice president became the project leader, and Kanaoka's factory manager, including the model line, from the union; Kitada, general secretary of the trade union headquarters; and Sakai, Yodogawa Shiga participated as members. These people were working as

FIGURE 2.18
Volkswagen's one-man system.

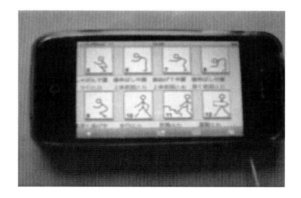

FIGURE 2.19
Cloud-type work posture measurement.

upper-level members, and the subordinates worked substantially under the model line of the worked the foreman class as a working group.

Professor Nagamachi was appointed as a consultant. This kind of appointment is very important if any organization wants to extend the outreach potential of its enterprises.[11]

Setting the medium-sized cooler workplace as a model line, an investigation was conducted on work posture, job design index, heavy goods handling, etc. An example of the results is shown in Figure 2.20.

All workers had complained of working posture discomfort in the elements of the "impossible." They also had problems in handling heavy goods. In reference to Figure 2.19, to get hard work to 5 points or less, we need to place the position of the shoulder from the navel to the workpiece (that is, to prevent bending the waist) by positioning the workpiece between the navel and the

Posture	A	B	C	D	E	F	G	H	I
Hardness index	1	3	4	5	5	5	6	6	10
Movement Content	Standing Posture	Standing Posture, stretch body, heel is floating	Slightly bend the hip and slightly bring forward upper body	Stretch the hip and slightly bring forward the upper body	A squat posture and heel are attached	Stretch the knee and forward the upper body half rising	Bend the knee half rising and forward the upper body	Stretch the knee and forward the upper body deeply	Deeply bend the knee in kneeling position and forward the upper body
Important Note	The upper body angle 0–30°	Posture where the item is raised above the worker's head	The angle is 0–50°	Angle of upper body 30–45° # over doing poster will look like (F)	Heel in floating condition (I)	The angle is 45–90°	Angle of Upper body 45–90° and the knee angle 0–45°	Angle of Upper body more than 90° and the knee is same level bended or touched	Heel in floating condition

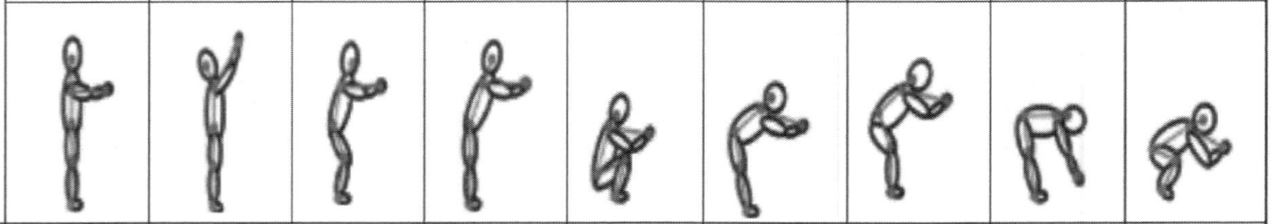

FIGURE 2.20

Nagamachi's cloud-type work posture distress index.

shoulder at all times, rather than bending over to suit the workpiece. This is a breakthrough idea. If you cannot maintain the position of the workpiece, let robots do the work.

Several improvements have been made in every process. We can see some process data in Figure 2.21. The figure of the fourth process comparatively shows bending at the waist, and both examples are shown in Figure 2.22; working on the long box that stands vertically (Figure 2.20) is a tense phsical process, but the platform where the worker is standing is moved up and down, he or she does not need to bend at the waist anymore. Figure 2.23 shows the situation when a cylinder is prepared under the box and then the cylinder arm is stretched and the workers are able to work in a standing position.

Although this line consists of 15 processes, only the result for the fourth process shows that the worker posture has changed, as in Figure 2.23. Although the posture became harder when the horizontal axis was moving to the left, there are fewer workers for N07 and N06 that have a difficult posture, compared to before the improvement process. Almost all the workers of N03 and below experience a comfortable posture as a result of the process improvement, where energy consumption has been reduced by about 20%. The result of implementing such an improvement on the total process for this line shows an overall productivity increase to 43%. Daikin is moving forward in the ergonomic improvements study in order to ensure the importance of workers' emotional health and as previously described, the production cell formula has also been successfully developed.

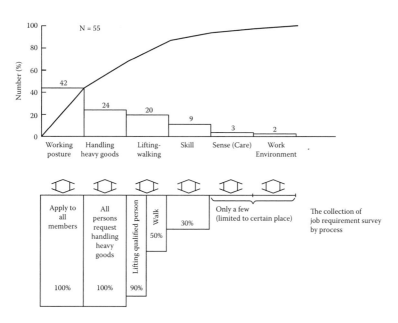

FIGURE 2.21
Survey of various worker postures.

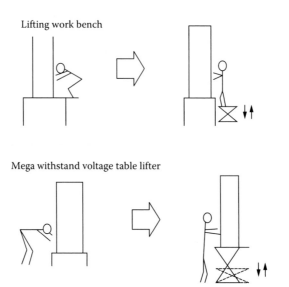

FIGURE 2.22
Improvement from relieving the worker posture (improvement from left to right).

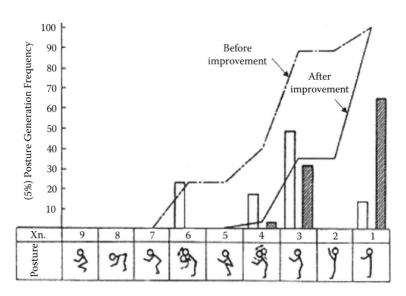

FIGURE 2.23
Comparison of before and after improvement for the fourth process.

References

1. M. Nagamachi. Application of participatory ergonomics through quality circle activity. In K. Noro and A. Imada (eds.), *Participatory Ergonomics*, 139–164 London: Taylor & Francis, 1991.
2. M. Nagamachi (editorial supervisor). *Textbook for redesign of jobs* (in Japanese). Association of Employment Development for Senior Citizens, Tokyo, 1985.
3. M. Nagamachi and Y. Matsubara. The ergonomic implication of a doorless system in an automobile assembly line. *Ergonomics*, 37(4), 611–622, 1994.
4. M. Nagamachi. *Job enrichment design—a system which creates job satisfaction* (in Japanese). Tokyo: Diamond, 1973.
5. L.E. Davis. The design of job. *Industrial Relations*, 6, 21–45, 1966.
6. L.E. Davis and J.C. Taylor (eds.). *Design of jobs*. Penguin Books, London. 1973.
7. M. Nagamachi. New design of jobs and ergonomics (in Japanese). *Ergonomics*, 9, 187–196, 1973.
8. M. Nagamachi. Design of jobs and motivation (in Japanese). *Journal of Japan Industrial Management Association*, 20(4), 302–307, 1976.
9. M. Nagamachi. Theory and practice of cell production system (in Japanese). *Management System*, 21(6), 299–308, 2012.
10. M. Nagamachi et al. Research related to development of improvement in the workplace using an ergonomics approach related to employee growth in the automotive industries and global improvement industry check system construction as a 2010 joint research annual report between the National Institute of Aging and Disability Employment Assistance Mechanisms in 2011, Tokyo.
11. M. Nagamachi et al. *Job redesign in Daikin Industries*. Employment of Older Persons Development Association, Tokyo. 1983.

3

Kansei Service Innovation

3.1 Organization Innovation in a Tenmaya Hiroshima Shop

Simply explained, service innovation is a "scientific idea" that aims to improve the productivity of the service industry by integrating an engineering idea. Other countries started with this concept long before, such as Germany and the United States. In Japan, the Ministry of Economy, Trade and Industry announced a policy in 2007.[1]

Professor Nagamachi has been fully involved in this area since 1969, when he started as a student, and achieved some great results from the stakeholders and employees. One day in that year, the general secretary of the trade union of the Tenmaya Hiroshima department store in Okayama requested him for consultation in the office of Hiroshima University. In this department store, managerial employees were also involved as labor union members, and they seemed to be worried about their employees' work motivation.

The main point of the consultation was to improve the employees' work motivation, including the elevator and escalator girl's enthusiasm, which was waning. Professor Nagamachi often visited the Tenmaya Hiroshima shop and felt that the girl's smile was mechanical, not sincerely coming from her heart, and that customers sensed this as well.

Professor Nagamachi spent a number of hours discussing behaviors and supporting the workers and supervisors. As mentioned in Chapter 2, the rules in working life include feeling work responsibility, self-development/self-growth, and a desire to produce a result that is highly valued by the company. Professor Nagamachi feels that the worker should not assume that the buyer is an expert on the product and has special knowledge, and that this store is the only store that sells the product that the buyer wants.

Professor Nagamachi wanted to give each member of the department store a chance to feel important. Therefore, it was recommended that the store execute serious training on product knowledge for all employees and supervisors and give them some responsibility for product purchasing, thinking that they would seriously sell the product they selected. The supervisor was happy and thought that it was a brilliant idea and returned to the office to discuss this with the branch manager and general manager.

The workers' union started a discussion internally, and thought that it was an interesting idea and agreed to try it; therefore, they entered the discussion with the management (officer) team. The objective of the discussion was actually a proposal to let all of the employees (excluding new employees who joined the company less than 1 year prior) conduct research on each product for its current attractiveness, raw material characteristics, sewing characteristics, cost structure, and saleable product characteristics, including the selection method, retail price and cost (profit margin), etc. Upon completing the study, the employees would be given a budget to travel all over Japan to offer the outstanding products at all counters.

The response from the company was that the Hiroshima shop had its own dedicated buyers for dedicated products, and to make the employees confident and specialized in product knowledge, they would need a couple of months of training. Because this would not result in any profit, the company rejected the proposal.

The union was confident of this idea and strongly hoped it could be executed. They were not giving up even though they received negative comments; Professor Nagamachi and Suwa, a fourth-year student (currently president of Kumahira Manufacturing), were sent to the Tenmatsu shop to experience the operations in the department store and conduct an awareness survey of the employees who would be involved in this project. There would be a risk if all the shops suddenly made changes; therefore, initially, just for discussion purposes, they appointed the chief secretary, who visited the university, to be responsible for the sixth floor's daily life goods (sweaters and underwear).

The guidance team consisted of the manager, assistant manager, and each product's veteran leader. After 3 months of intensive training and receiving a proposal from each employee that had a budget for purchasing goods, they visited each location (store in Ginza Tokyo, general store at Kyoto, and Hiroshima's famous wholesale). This new method was called the self-order system. This is a system where the ordering is done by the employees.

An explanation was given by the members returning from purchasing to the members at the counter, and each of them understood the overall message and conducted collaborative work by adding attractive words to the showcase content, product arrangement, and the detail and good point of purchase advertising (POPA). The next day, they transformed the selected products for the new showcases selected. The workers' faces shined at the opening of the shop at 10:00 a.m. Most of the customers enjoyed being welcomed by the worker with a loud voice, which they had never experienced before, and also, some of the customers who had come before noticed that now the worker was explaining the added value of the product with a confident voice. The sales are also improving gradually. The largest improvement is the members' motivation and job satisfaction.[2]

As a comparative study, the second floor, which had a similar product with a high price was considered. The responsible management had not been introduced to the program, but was given the same questionnaire (51 items) on the same day for comparison. The employees' motivation was very high for the sixth floor employees, and the management communication level was also frequent for the sixth floor. The target value (job reward) against display ability, approval, and sense of accomplishment greatly contributed, and the result was consistent with the Herzberg motivator factor. Sales also increased more than the year before.

The demand from other departments to introduce the suppliers' liability system, which started on the sixth floor, gradually increased, except for the dry goods and jewelry areas. It also spread to other departments, but after 3–5 years, the management closed it down and finally discontinued it, except for the necktie section.

3.2 Shop Master System at Seibu Department Store

Professor Nagamachi describes a story that caught his attention in a book published by Diamond Publisher, where Seibu Department Store was supposed to perform a transformation.[2] His idea was that inside a small shop in this department store, a young leader was to be in charge for all management: employment of subordinates, training, purchasing of goods, and profit management. Then the group would become highly motivated as an autonomous team. Additionally, if some of the profit were divided among the team members, their satisfaction would be greatly increased. Professor Nagamachi named this idea the "Shop Master System." This idea is better, theoretically, than the supplier's accountability system at the Tenmaya Hiroshima shop.

In the beginning, a number of stores that launched the shop master system suddenly increased to 80%, and this system suddenly widened to all the stores in the department store, excluding the cinema and specialized store. The Seibu Department Store was the head and each department store formed a series that spread to all the inside shops within the Seibu Department Store, as well as all the children department store subsidiaries under Seibu. The Takashima series also expanded to other series, and then the Tenmaya department store also introduced it. When Professor Nagamachi visited the Tenmaya Hiroshima shop again for shopping, the managers were much happier and said to him, "We are truly sorry now." rendering him speechless. A good system brings good effect. Sales are also growing, and job satisfaction and the morale of employees have improved significantly. This kind of example is called "human engineering world job design" (organization design).

3.3 Kagaya Service Innovation

Professor Nagamachi believes everybody will think of Wakura Hot Spring (Kagaya) when we talk about the best Japanese service hotel in Japan. In December 2010, when the Taiwan Kagaya Beito Hot Spring opened, it received delegates from Taiwan. Very skilled and possessing great hospitality, Kagaya became a world-class hotel. Kagaya was founded in 1906 by Oda Yoshiro in Noto Peninsular (Figure 3.1).

It is well known that the Showa emperor, empress, and family were customers of Kagaya. Kagaya at that time consisted of 246 rooms (capacity: 1460 people) and was supported by 650 employees; as a business, it was continuously prosperous and believed it should construct a new facility in a separate building to meet its popularity.

At this time, CEO Mr. Oda thought that the best service was to provide room service to each customer. However, as the service women aged, carrying food trays became heavy work for them. He asked for a consultation by the office Foundation for Older Persons Employment Development (Tokyo).

Professor Nagamachi has found that excellent service to make customers happy is for the elders like Nakai-san (skillful service lady), who had a good technique, but not for young employees. Oda, the president, concluded that the hotel could not lower the hospitality level, but for the old Nakai-san, it could not improve the work anymore.

The issue to resolve concerned balancing the actions of the workers carrying the cuisine and the excellent hospitality required. Dealing with the Kataoka manager, the managers discussed details of workers' issues, visited the site several times, and finally achieved a great solution.

FIGURE 3.1
Hotel Kagaya hospitality. (From home page.)

FIGURE 3.2
Kaga Shop food automated transport system.

The solution was to construct an automated transportation system with robot food carriers to serve each customer room. When the food was ready and still hot, the robot transported it to the customer room, where it was delivered at the specified time. The Nakai-san then served the food with the appropriate hospitality.

Although the transportation was done by a robot, the hospitality skill was by Nakai-san, and the service was further improved by sharing both skills. In other words, the problem was solved by introducing innovation into the service industry (Figure 3.2).[3]

References

1. Ministry of Economy, Trade and Industry. *Service industry and service innovation towards the productivity improvement.* Tokyo: Economic Industrial Research Association, 2007.
2. M. Nagamachi. *Design of Job Enrichment.* Tokyo: Diamond Publishing, 1973.
3. M. Nagamachi. *Challenge of comfortable factory—ergofactory.* Tokyo: Japan Maintenance Association, 1996.

4

Kansei and Organizational Management

4.1 Suntory Amoeba Organization

The term *amoeba management* is known in Kyoto Ceramic (currently known as Kyocera), founded in 1959, and launched by Kazuo Inamori. It is a mechanism of corporate organization or organizational management. Fifteen years later, Suntory, as an efficient management organization, ends up with an amoeba management formula, but using another appropriate term: the self-management method.

With the objective to enhance the whisky supplies in western Japan, Suntory planned to establish a factory in Miyajima of Ono-Machi, Hiroshima Prefecture, in 1973 and appointed Kishimoto, who graduated from Hiroshima University's Engineering Department, as the construction groundwork leader. The headquarters' human resources department proposed preparing 27 workers based on the size of the plant and taking support into consideration. On the other hand, Kishimoto calculated that 72 workers were needed, which resulted in a conflict with the human resources department.

Kishimoto asked President Saji to increase the number of workers, but the president decided instead to construct a factory in Miyajima using 54 people. Kishimoto kept thinking about the president and the human resources department suggesting execution with a small number of people, from the setup preparation to the expected workers; the workers were required to have multiple skills and had to get cooperation from subordinates.

In 1974, the factory accomplished having one person having four skills. As shown in Figure 4.1, for example, person A can do electrical repair, quality control, be in charge of the whisky sensory check, and also be in charge of accounting. The time is allocated, where there is a time that he or she does the work alone at the site, and there is a short time where he or she sits at office and does the calculations of other workers' outstation processing.

Each employee had not three but four job responsibilities. This was in preparation for Kyoto where they analyzed everyone's duties and, whenever possible, transferred part of the duties to the computer. The duties transferred to the computer were also quickly executed at the new plant.

FIGURE 4.1
Santori Miyashima Factory multiskills.

This system was already programmed before leaving Kyoto, where all jobs were broken down into small tasks, and, whenever possible, parts were transferred to the computer.

This kind of improvement allows incorporating small group activities, where each group will analyze its duties and use its knowledge to computerize them. For example, for the general affairs section in an office, a receptionist, general affairs, accounting, and the person in charge are defined, and each of them needs a workplace. At Suntory, the desk is a round table, and does not include a drawer.

When employees have a drawer, they put things inside the drawer, including work that must be done, and it becomes someone's desk. Anyone can sit down at any seat in an empty place. When there is no drawer, the work is finished collectively by everyone, documents are stored in the locker on the sidewall, and nothing remains on the desk. Improvement work is carried out without hesitation, and people are constantly looking for work to be completed. The next year, the staff was reduced from 54 persons to 43. It was targeted to be reduced to 27 persons in 1976, and finally to 19 persons in 1980. On the other hand, the production output was increased two to three times, and output volume per person was also significantly improved several times by the forecasted plan.

This phenomenon is called the *amoeba system*. It is an activity or lifestyle where a single amoeba (worker) tries to solve a problem or challenge.

During the gift-giving time in summer/winter, on Saturday/Sunday of the holidays, all the workers visit the nearest departments and talk to the salespersons discussing the customers' needs.

Even though there are periods when they cannot reduce the number of workers, reallocation of duties can occur. For example, let's say there is a small group of five excellent workers. If someone shouted, "Tanaka passed away," on the spot, the workers would make a detailed breakdown of Tanaka's job and share or divide those duties among other employees. Tanaka, who is no longer present, indirectly says to Kishimoto, "I am no longer available. Please make arrangements for my duties to be reallocated."

All these things happen in harmony and calmly in the organization, in which the worker's Kansei and their full cooperation were unexpected outcomes. The small group activation is a great mechanism in the development of values and sensibility.

4.2 The Idea of a Mini Company

The mini company was established by Sony and led by executive director Kobayashi, and involves small group activities. As previously described, Kyocera uses this concept along with self-management activities. The next well-known mini companies are the Nippon Electronics (NEC) Tamachi factory and another company, Maekawa Seisakusho. In the Tamachi factory of NEC, the manufacturing site is divided into several unit groups, and each works as a mini company.

Interestingly, a group consists of a president, vice president, production manager, quality control manager, labor manager, and other executives who volunteer, but the decision is made by approval of all the members.

For example, someone will say, "This season I will be the president. I will reduce the cost price by 15% and improve the profit." The group members will know his or her personality and skills well, and will have approval on whether to select him or her as president.

The selection of other executives also requires a majority vote opinion in support of the candidate. Most of the members will have their role, and that is the NEC image. Thus, at the completion of the first period, each group calculates its production outcome and sales value, and evaluates its performance.

Continuing in the second period, workers, including the selected officer, participate in a review meeting. Whoever has not considerably fulfilled his or her performance obligations the previous fiscal year will not be approved at this time. In this way, performance is always improved, and the operation continues to run by the voluntary method. Even young people are allowed to become candidates for president.

After the decision has been made, the entire membership will cooperate with each other to improve their efforts in order to produce an actual product. It is not acceptable to express negative comments about the president, and taking advantage of the naiveté of a younger presidential candidate is also unacceptable. Small group unity is difficult to achieve, but if the group members are not performing, their bonuses will not be affected.

In general, harmony is when you have cooperation, no arguing, and different opinions are dealt with. If the performance is improved, the same president will take leadership for years.

Maekawa Manufacturing, founded in 1924, well-established by Kisaku Maekawa from the Maekawa store, is an ice making from cold storage operation for various refrigerator manufacturers and various compressor development manufacturers (currently Japanese manufacturers and international manufacturers). It has expanded to companies that have a sales office.

The second-generation president graduated from Waseda University in engineering management and is a respectable person. Here is a whole company that consists of each unit functioning as a mini company. For example, in the manufacturing sector, there is a small type of compressor department, a medium type of compressor department, a huge type of compressor department, a research center, a quality control department, and an indirect department, and each of them is an independent mini company.

In the factory, each section is similar to NEC, where each executive is selected by self-recommendation and overall performance, the same as a normal company in the office or laboratory. Generally, the difference with the company is that each mini company has self-supporting business activities.

For example, the Hiroshima sales office will keep in mind that it needs to turn a profit, so its products should have even better quality and cheaper price, and it is allowed to sell other companies' products (third-party products).

The research center has to develop innovative products, and the office also allows it to use third-party products. This is called the "collection of rivals" by the Maekawa factory, where each seeks profit for its activities.

Ultimately, all divisions need to perform activities directed by the sales department, but the business may not be useful, although the president volunteers himself and needs approval from others, it is a sincere decision.

4.3 The Birth of Motivation at Mitsubishi Electric Mechanism

In the previously called Mitsubishi Electric Co., Ltd. Nakatsugawa factory, Abeyama is promoted from production technical manager to production manager, and the one-layer cell production system becomes more actively used, and all lines in the factory are introduced to the production system.

In the Iida factory, the newly set up substructure of the factory mainly combines home appliance fans. As the Iida factory opens, its parts are carried by a conveyor belt, and each worker picks up parts from the conveyor cell line. Each cell type consists of 20 female workers, and although the parts use a belt conveyor, each person is good with each part (i.e., it is not of the conveyor takt type) and assembles all the parts with her hands.

The completed fan is placed at the lower part of the belt using a two-step formula and sent to the final inspector. If the inspection finds no problem, the completed product is sent to storage.

This means that the cell type consists of 20 people gathered, and the belt is just a transport belt and is not made in the normal takt system where each person independently does the assembly process.

To avoid assembly mistakes or poor quality, the 20 workers have different color magnets that represent individual assembled product installations, so the inspector will know instantly where a mistake happened. Then the inspector goes to the operator and explains the cause of the mistake and asks for correction.

Because it is impossible to achieve zero mistakes, Professor Nagamachi suggested that one of the 20 workers be assigned to be in charge of inspection in a one-week rotation. By understanding other persons' mistakes, zero mistakes should be achieved.

While group leaders were conducting detailed training on the quality management system, it was noticed that when a worker found a defective part, he or she always gave the part to the group leader to pass on to the quality person in charge. After training, when a worker finds a defective part, he or she calls the quality control manager or subcontract manager directly and gives guidance. As a result, the quality problem is solved and the reoccurrence of the problem is negligible.

This kind of activity is possible and uses the cell method production independently, where the workers can discuss with their supervisor, and overall the production problem will decrease.

In a different story, when the operation is started, the drinks, cookies, and stationery shop are opened. From the beginning, without interaction with the customer, the cashier would calculate the coins and put them into the register. However, they also had a hard time when coins were not matched, but when the cell system of production improved, the mistake of the calculation was eliminated, whether the workers put no change in the cash register or there was an excess. This is the result when autonomous-production and self-action are practiced.

The very interesting part is when the production volume is planned monthly and dictated as a goal by the plant manager. Although it is not easy, workers manage to achieve the goal it 1–2 days before the last day of each month. Using the sale production system, workers are united and try to finish as soon as possible. The extra time is used for cleaning, machine maintenance, or asking the supervisor to open a class to explain the root cause of the latest defect occurrence. This entire thing is not managed by the plant manager, but the workers are doing it based on their own initiative.

Why does this kind of voluntary action take place without any pressure? Because workers feel satisfied and have a sense of job satisfaction. Let's think of a normal factory. In a big factory, there are many lines and many workers moving around, and although the entire monthly target is displayed, when a worker is asked, he or she cannot state the target. The things that become more and more compared to the previous month are the mistakes and resultant line stops, which are always everyone's first thought.

However, we can think of the mini company as a small number of people in a group, and each person has his or her own targets and responsibilities, and this will become the short-term target. In addition, the responsibilities associated with the target are reasonable. When we talk to everyone, they are concerned with their values and also sensitivity. Since the target is set by discussion among them and is not considered a command, it is the group performance that works toward the target. It is the same thing in a mini production system. In this case, if they divided the work among all, the goal or target for each person would be determined. If every person tries to produce better output, it is based on his or her own responsibilities and goals. The goal/target can be achieved if there is a team spirit in the group—unless they became overconfident. In conclusion, such a voluntary production system makes job sensitivity work, and work is equal and psychologically creates job satisfaction and self-motivation.

4.4 Result Produced from Small Group Activities

The mechanism of the mini company is recognized for enhancing organizational power, which strengthens the unity of the group. The same thing, the strength of the organization, is the center of the workplace (small group activity).

In academics, the University of Michigan has focused on the principle of the small group activity in its group dynamics research. In practice, many organizations receive guidance from the work of Dr. Deming as a method of quality control, and the Japanese have long advocated group activities.

Although Japan is practicing small group activities and producing organization outcomes, when we talk about other countries, basically the Japanese culture is criticized when it tries to introduce this because it is viewed as not being practical. And as described, in the world oil and gas industry, Chevron uses the power of small group activities, and holiday accidents are almost zero. In 1987, the new setup of the Malcolm Baldrige Award focused on small group activities and leadership.

There is a procedure for small group activity establishment (Figure 4.2). Initially, group activity begins informally. Unlike others, Japanese people are not good at talking. We don't know what to talk about. Also, there is

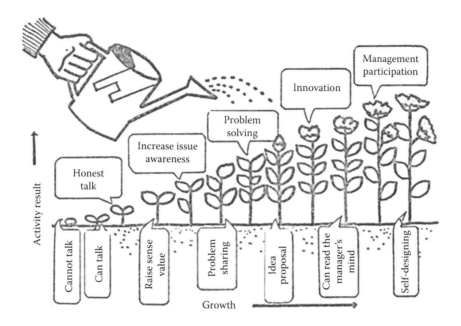

FIGURE 4.2
The growth process of small group activities.

a time when the topic is shifted in a different direction. We begin with the fact that the senior employees and the group leader are not good at talking, and we guide them gently. First, the discussion is started with an everyday topic, especially for the young workers; they are asked questions such as "What do you think of this?" When the discussion has been gradually established, let them consider a particular issue. The first step is to open discussion.

The second stage after getting people to talk is to start with problems at their workplace. They start to realize the problems and think of problems that can occur. The group will talk about the problems, and everyone understands and seeks direction on how to solve the problems. The second stage is called the awareness stage. At this stage, everyone thinks about how to resolve the problem, and tries to positively move forward without causing trouble in their workplace. When everyone is involved in solving the problem based on their ideas, the process becomes interesting and improves the workers' problem-solving abilities. The third stage is the actual problem-solving stage.[1]

At this stage, understanding the needs of the company is important, rather than receiving instructions from a superior, and based on that direction, they will understand that problem solving will have a positive impact on everyone. Thus, we will reach to the fourth stage of creative solution. The ability of the group will grow productively through each stage.

On reaching this stage, they will think to move the company in the desired direction. Naturally, everyone tries to understand the issues and challenges, and acts as supervisors and management in actively completing the problem-solving exercise. That is exactly what happened to the female workers in Mitsubishi Electric, Iida factory.

When one of the groups in Nissan Diesel achieved a great improvement, other groups learned the technique and also achieved great improvement. Then, the unit that originally made the improvement was painted in a purple color to represent the improvement implemented. After a while, the purple color was seen in more and more units, which motivated everyone to make improvements so that their units could also be painted.

As a result, the entire workplace has clearly had improvements in production, lower costs, and has transformed to a workplace with good production efficiency.

The term for this group activity, which internationally is called *participatory ergonomics*, was also created by the Japanese, which Professor Nagamachi said was like fate. Participatory ergonomics is when members are involved in the problem, and sometimes it will involve the management and can be solved by using the knowledge of human engineering. In this way, the small group activity belongs to the members' sensitivity and ability to understand each other and grow with values and feelings. The Japanese are raised with this shared awareness and have this ability.

4.5 Initiation of Hakuryuko Country Club

While preparing to guide a production process for a primary cooperative company that supplied bumpers to Mazda Motor Corporation, such as Daikyo Co., Ltd., Professor Nagamachi received a request from the owner, the president, who said to him, "Although I have a golf course, I cannot earn a profit. Could you advise or guide me?" Professor Nagamachi is totally not good at golf, but he can give guidance related to management. First, Professor Nagamachi met the manager, who guided him to the surroundings of the golf course, and listened to the current problems at the Hakuryuko Country Club. He was informed that employee motivation was weak, there were fewer customers from Hiroshima District, and customer relations were also not good. Initially, Professor Nagamachi recommended giving guidance on small group activities.

Hakuryuko Country Club is located near a large lake called the Hakuryuko, and it is a nice place for a golf course. However, it is difficult to retain popularity because the location takes about a 1-hour journey from Hiroshima, and Hiroshima already has many excellent golf courses (Figure 4.3).

FIGURE 4.3
Hakuryuko Country Club.

All the employees were gathered in one place and divided into groups of five or six people, and the procedure of small group activities was explained, where they could freely talk and, if possible, share good ideas, and most importantly, discuss how to please the customer.

Initially they complained about company, but gradually the problem became clear. Interestingly, one of the middle-age females said, "To get more profit, everyone should do the work of two persons." In the early morning, customers would rush to the front, and the receptionist would be very busy. When the customers went to the course, the receptionist had free time. So the following suggestions from Professor Nagamachi were applied.

The receptionist does her up-front desk job in the morning; when the group goes out to play, the receptionist caddies for the same customers and is responsible for the golf clubs. When half of the play is finished, the customers rest at the restaurant. At that time, the caddy becomes a waitress, wears an apron, and takes food orders from the group. Because this group remains together, they will become friends with the caddy, which makes the ordering process run smoothly. This proposal was mostly agreed upon by the female workers. They thought it was interesting and started to discuss the mechanism from the next day.

On the next day, each person started with three job roles. The front receptionist left with the group and became a caddy to assist them. When half of the play finished, she took off her uniform and wore an apron and waited at the restaurant while the customers had their cigarettes.

When all the players in a group came into the restaurant, saying, "Hi, Caddy-san, what is your recommended meal for today?" The caddy, or waitress, would say, "I heard that we just received good quality meat today.

It is a delicious and an affordable price," even if they recommend the cuisine that is slightly higher priced. The order is confirmed immediately. Because of that, the restaurant sales increased immediately. In response to this activity, the greens management group and the bath management group have been actively in transformation, unlike previous times.

For example, according to the bath person in charge, the number of towels used in the bathroom were two to three times more than the number of visitors, and as a result, the number of towels and the cost of detergent also increased. While planning for this improvement, one of the members shared an idea: by folding each towel in half, like tissue paper, customers will naturally take one piece each when they want to use it, and the usage of towels and detergent will be cut in half from the current situation.

Also, during a round of golf, a player who is concerned about the distance from the green to the hole might ask the caddy for an opinion. Knowing that players want this kind of information, the caddy group could go out on the course during a holiday when there are no customers, and measure the distance to the greens, the trees, and the stone, and note the different kinds of grass. The club's reputation improved because customer scores improved based on the accurate distance information given by caddies who had memorized them.

Business improved, visitors from other provinces increased, and the club was reevaluated from a three-star to a first-star country club. This scenario is considered an example of successful small group activities, and the workers participated in the Japan Science and Technology Federation QC Circle Tournament and won the top prize of the year. Since that time, other golf businesses around the country visit this club for reference.

Reference

1. M. Nagamachi. *The psychology of QC circle*. Tokyo: Kaibundo Publisher, 1987.

5

Activating Kansei in Product Development

5.1 Activating Kansei When Creating a Product

Because Professor Nagamachi was as an assistant professor prior to his work in quality control, he was strongly influenced by two professors from University of Hiroshima involved in the Deming Prize jury. His specialization was in total quality control (TQC). The quality control outline is getting the customer request, starting with the customer need, setting the guideline, and educating product management to maintain quality by changing unbalanced activity according to customer needs, in other words implementing Kansei engineering.

Beginning in the 1970s, the term *emotional engineering* was already in use, but most of Professor Nagamachi's foreign colleagues at international conferences used *emotion* in the context of bringing strong emotions like grief and pleasure. So, Nagamachi was advised to use the more suitable Japanese term: *Kansei*. Then in 1986, Professor Nagamachi received information from friends at the University of Michigan that Kenichi Yamamoto, president of Mazda, gave a special lecture titled "Kansei Engineering" at the University of Michigan; thus, the term *Kansei engineering* came into use, and now is known the world over.

Kansei is a word the Japanese would normally use to express sentiments such as "I wish there were something like this," "I want to do this," "beautiful," "delicious," and so forth, which are synonymous with emotion. When a customer makes complaints such as "Why do I have to do this?" or "I wonder if this is sold at the supermarket," it is time to develop a product to fill that need.

When you want to do something or feel inconvenienced, you feel an emotion; for example, when you feel something such as "delicious," you will directly react with emotion. This kind of emotion or sense is grasped in advance. What you can express is taste, shape, design, etc., and if you design a product in an appropriate way based on customer emotion, you should be able sell it. Product development is the most active use of Kansei.

To start creating saleable products through Kansei, it is important to accurately capture the emotions of the customer. As discussed in Chapter 1, there are various forms of emotion that can be expressed. Based on that,

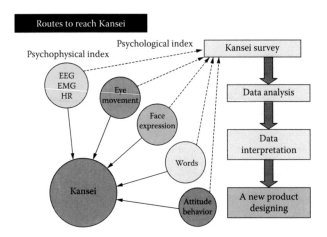

FIGURE 5.1
Route to reach Kansei (method).

the most accurate way to capture emotion easily is to use the selection method. Refer to Figure 5.1.

The center for Kansei appears in the brain's physiological content in various formats, such as brain waves, functional magnetic resonance imaging (fMRI), Nuclear Information and Resource Service (NIRS), electromyography from the body, galvanic skin response (GSR), etc. When you have an interest, open your eyes and focus on the target.

Your expression will change when you eat delicious food, and you will say, "Delicious!" Kansei depends on the thing on which the exposure of emotion is based, and the type of emotion could be different for different people. It is important to consistently understand the method, where the goal is the design. To do this, we need to collect Kansei data and analyze these data using physiological, engineering, and statistical analyses to produce useful information.

These kinds of analyses reflect only the current state information. We use this information to develop new products to satisfy customer needs expressed through the Kansei data. We continue to gather and analyze Kansei data over time to determine our progress and realize future product development. This is the ability of Kansei engineering.

5.2 Recommend Customers Touch with Hands

There are two points to consider when performing product development and utilizing customer emotion. First, customers feel attracted when they see a new product and then they touch it and check whether it impacts their feelings or not. When they don't feel an impact when they touch it, this means that the opportunity to get in touch with the customer is not developed and there is no chance of the product being purchased. We cannot call this a new product. Attractive elements of product appearance include shape and color that attract the eye to the appearance and hold an attractiveness characteristic. You will get a feeling such as "Oh, there is something unusual. I wonder what it is." Certainly, this characteristic of attractiveness is the result of Kansei analysis and must be taken into account in design.

Second, when you feel the attraction, then you also feel the impulse to buy and try the product, and unexpectedly your face is shining. This opportunity can be repeated, and is good if you can realize it. Feeling attraction in the face is called *primary first moment of truth* (FMOT). It is called *secondary FMOT* when the attraction comes after the purchase. It is a great success if you can design new products that have this characteristic (Figure 5.2).

According to the Kansei research so far, the first element to create attraction (attractiveness) is the structure of color, which represents about 70–80% of attraction. The second element is the shape, and furthermore the roundness design part is greatly affected. You can get the weight of a factor by using Kansei analysis and need to consider when to proceed with the analysis.

Google recently emphasized the importance of the zero moment of truth (ZMOT). The zero moment is actually before we see the real thing in person, where the customer somehow received the information, via TV, Facebook, Twitter, etc., and already had the intention to purchase or not. This kind of

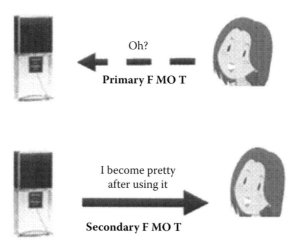

FIGURE 5.2
The changes from attraction to purchase to repeater.

media information creates the attractiveness and the effective opportunity. We need to use Kansei analysis from media information for advertising. Professor Nagamachi introduced some new Kansei products that were developed under his responsibility.

5.3 Human Living System

The human living system is the first to apply Kansei engineering. It was broadcasted live at 7:00 a.m. in Tokyo NHK Studio 103 in 1974. At that time, Professor Nagamachi and a female student discussed and explained the use of Kansei engineering in the interior of a live studio. There were many callers from the NHK staff and external callers from the audience or viewers that wanted to make use of Kansei engineering. The company was NishiNihon Interior Co., Ltd. in Fukuoka.

Two years earlier, Iwashige Ritsuko, a master student of architecture, applied Kansei engineering in her research thesis in the study of interior design. She produced a human laboratory in a 1/10 ratio model with walls, floor, curtains, etc., that could change color in several ways and let students evaluate the Kansei. When analyzed through factor analysis, the results reflect the emotions of the specific person to a particular interior decoration rule. The name of her new product was the *human living system*.[1,2]

The customer visits the company when he or she wants to purchase a room in a new building or apartment. The salesperson can easily explain Kansei engineering by showing a pamphlet of the human living system. And then the salesperson presents the video created by the Hakuhodo. They describe each person's feeling of staying in the room using their own words.

For example, "comfortable," "luxurious," "Western style," "I will be in a relaxed mood," "It makes me want to listen to music," "It makes me want to talk to my friends," etc. All these are Kansei word expressions. An actual example is shown in Figure 5.3. This is one of the examples that comes from the same Kansei words. When a Kansei word is identified, the product code combination of the walls, floor, and ceiling of the room and curtains is identified. Each product has different colors, as reflected in Figure 5.4, which has a set of five combinations based on the same emotion, and customers will select their favorite color from them. For example, a customer said, "When I enter the room, it makes me want to study." Corresponding to that, the salesperson will select a suitable atmosphere for studying. A customer who is a child can find a suitable atmosphere from one set.

In a short time, the profit was about 2 billion yen, as all the newspapers such as "Nihon Keizai," "Nihon Kogyou," "Asahi," and "Yomiuri," ran reports of the system. This company was also fortunate in that it had worked on the interior decoration of Mitsui Home Co., Ltd. This mechanism was used for

FIGURE 5.3
Luxurious example of Kansei.

FIGURE 5.4
Selection of showroom walls, floor, ceiling, curtains, etc.

about 200 types of material, all of which were purchased from Sangetsu (Ltd.), but from Sangetsu's perspective, it could only sell 200 types without orders from other companies.

Without giving his name, Professor Nagamachi visited Nagoya Sangetsu to learn the real circumstances in Sangetsu, and, as compensation, he gave advice on the method of showroom sample display and went home without being recognized. Of course now, Sangetsu is back to being a key player and the top company in their industry.

5.4 New Refrigerator Development at Sharp

In 1978, Professor Nagamachi received a request from the design executive in charge at Sharp asking for guidance on Kansei engineering for its 150 designers. The designers were all gathered at the Sharp headquarters, and although the explanation was conducted twice, they complained it was difficult to understand. So Sharp decided to apply Kansei engineering to a new type of refrigerator and hired Nagamachi to develop a new type with two doors.

To do the new thing based on industrial engineering, Nagamachi decided to observe and monitor how a housewife uses a two-door refrigerator each day. He organized a team consisting of six female designers and monitored these homes with a camcorder.

Since two-door refrigerators were common in those days, the camcorder was installed in front of the refrigerator and continued shooting for 2 hours while a meal was cooked. When the scheduled investigation finished, the members were asked whether they found any clues during the observation, but they could not come to a conclusion, so Professor Nagamachi decided to look at the video recording again.

Some people were asked to record the number of openings and closings for the upper and lower doors, as well as the main item that was taken out. These data are important to help us think further. Even with six people monitoring the recording, they could not come up with an idea of what the new refrigerator should be.

In those days, it was normal for the upper door of the refrigerator to be used for food freezing and ice making, while the lower door was used for the dish to be prepared that day. When Professor Nagamachi looked at the video, he found that the lower door opening and closing represented 80% of both door opening and closing. Most instances were for vegetables, tofu, eggs, and others materials intended for use in cooking immediately. In brief, the result of this investigation concluded that the homemaker was always bending down to open the lower door to take out vegetables, etc.

Here, Professor Nagamachi recapped the discussion in Chapter 2 on the problem of working posture. In order to open the lower door of the refrigerator to take out vegetables, for example, the person needed to bend his or her knees and hips,[5] using four to five times the energy used in a standing position. Homemakers were always in a difficult posture but did not complaint about it, so the engineer and designer, who were assigned to study the emotional aspect, needed to take care of the bending and stretching posture even though it was not complained about by the customers.

As a collaborative work, the new product development group and designer group changed the lower door position to the upper door position, and the upper door position to the lower door position, and the difficult posture was avoided as shown in Figure 5.5. This has become the new design. By the way, to make customers even happier, an investigation on what they stock in the refrigerator was done and a mechanism invented to allow customers to control the temperature corresponding to the storage of each food.

The upper compartment of the refrigerator is the storage compartment for tofu, eggs, beer, etc., with 2–5°C. The lower compartment is divided into two storage parts; the freezer compartment contains upper and lower storage, so that the temperature can be adjusted to coordinate with the different frozen foods. This temperature control device is the new design. In the video, a person removed frozen meat, put it on a table, and then went out, allowing it to retain its taste. Defrosting temperatures were estimated and a chilled area was invented and placed between the freezer and the refrigerator compartments. With the range of 0–2°C, it maintains the taste of meat and fish.

With these improvements, homemakers no longer needed to bend frequently at their waist. Kansei justified the development of the new refrigerator.

Starting in 1979, Sharp manufactured the new type of refrigerator for a while, but soon after Hitachi, Toshiba, Matsushita, and others started manufacturing it too. Of course, at that time Sharp was the top maker of the refrigerator. Thereafter, this model received a high reputation in Japan for a long time. Recently, we discovered that Samsung from Korea sells exactly

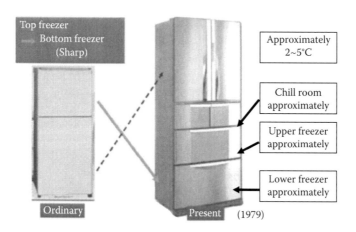

FIGURE 5.5
From two doors to a new type of refrigerator.

the same type of refrigerator, the concept of which had been duplicated from this. It is popular in South America and Southeast Asia as well.

There is great interest for other parties to see the activities of Kansei development and implement it. This group developed what Professor Nagamachi did during the refrigerator development activities related to the video shooting (camcorder).

The market share of Sharp at the time was about 3%, and management was thinking of stopping the production. The group asked permission from management to analyze how to use the development process for the refrigerator as well as the video shooting device that was used to shoot at the site.

A perfect opportunity to purchase the video shooting device is when a child is born. A young father is able to video his new baby's face, and as the baby grows, he can videorecord the baby crawling on the floor, which was previously impossible to do. The focus of the development was based on this need, and then a lens was invented that can be rotated 350°, so that the user does not have to assume an impossible position.

In the topmost part of Figure 5.6 is the traditional heavy camcorder. The second image is the innovative recording equipment with an image storage system and a liquid crystal, which is the LCD ViewCam, and slowly it became smaller and eventually evolved into a digital camera. Currently, every smart phone is built with a camera installed. The evolution until now is shown in Figure 5.6.

5.5 Wacoal Good-Up Bra

Wacoal sold a bra (Good-Up Bra) that, at that time, became popular not only among teenagers, but also among young women. It brought a new era to bra design for other companies to follow.

Wacoal manufactures underwear for females. It not only sells in Japan, but is also known internationally, and its development ability is very highly regarded. Professor Nagamachi received a request from Nakano Hiroshi, the production quality development executive, to work together on introducing Kansei engineering in an attempt to develop a new bra product.

When Professor Nagamachi finished his lecture overview of Kansei engineering to the employees, including the designer, he received excellent development and research capabilities from the members of the Human Science Research Center and proceeded to the operation.

First, Professor Nagamachi started to understand the emotion of using the bra from the female user's perspective. In surveys of 2000 women regarding emotions and feelings, questions such as "What do you want to achieve when you wear a bra?" were included. Eighty percent of the females answered that they are wearing a bra in order to be "beautiful"

FIGURE 5.6
From the traditional type of camera to digital cameras and LCD ViewCams.

and "elegant." In other words, "I want to be beautiful and elegant" is the need of the female, and if he achieved development of a bra based on this need, the female customer would be satisfied and would buy the product.

As part of the second phase of the Kansei engineering research, Professor Nagamachi conducted the most popular research and analysis. He collected all types of bras from various makers and of various sizes for young females to try on; each was evaluated by these two Kansei words: *beautiful* and *elegant*. He created a score list to evaluate the collected bras using a scale of 10 points, a product is beautiful = X points, a product is elegant = Y points, etc.

Next, Professor Nagamachi examined the collected bras one by one. The material structure, size, and shape of the cup, fiber-expandable characteristic, diameter of the under wire, color, and design, etc., were recorded on the list by examining all the features in detail. Then he established the Kansei physical characteristics of *beautiful* and *elegant* using a combination of the Kansei score by the tester and the evaluation of physical characteristics of the product, and performed a multivariate analysis. He used the physical characteristics to produce the new bra.

Next, the completed new bra was modeled accordingly as designed from the data interpretation. Using the Moire scale developed by a professor from the Department of Engineering at Shizuoka University, Professor Nagamachi obtained, by imaging, the unevenness of the body when a female was clothed in the new bra (Figure 5.7). This imaging provides a better understanding of the irregularity of the body because it is a contour plot. On the left is the traditional bra, and on the right is the Moire measurement diagram for wearing the new bra.

The design principles for the new product were that the left and right breasts needed to fit inside the left and right body lines, and both breasts needed to be aligned parallel, facing somewhat upward. To understand this, see Figure 5.8, where the left side is from a traditional bra in which the breast appears with impropriety and the left and right breasts face each other in the opposite direction (undesireable).

With this, the new product, using the Kansei experiment and analysis, was validated to create a product that fulfilled the female's needs.

With the Wacoal bra, both breasts are in a raised forward shape; this new product was named Good-Up Bra. It was very popular, with young females purchasing three or four bras, and brought the sales to 750 million yen. This happened in 1992. Of course, other companies also followed, but they could not catch up due to the different design concepts.

Wacoal created a second edition of the Good-Up Bra, and created sexier products such as the Hip-Up Panty, which was continuously a hit in the market.

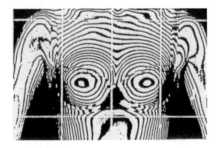

FIGURE 5.7
Comparison of the traditional bra (left) and the new Good-Up Bra (right).

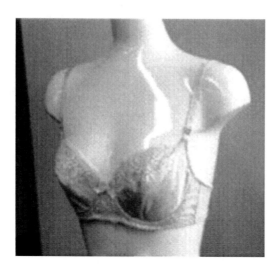

FIGURE 5.8
Good-Up Bra.

5.6 Hit Product in Komatsu

The Aoki managing director from the giant construction company Komatsu came all the way to Hiroshima and requested that Kansei engineering be applied at Komatsu. Immediately, Professor Nagamachi visited the construction machinery training center in Hakone (near Tokyo), examined the current products at Komatsu, and experienced them, such as driving a large dump truck, to understand the basic concept.

This is not the first experience of Professor Nagamachi in this area, as he also had experience as an ergonomics researcher at the University of Michigan Transportation Research Institute from 1967. At the University of Michigan, there is a 4000-m runway, where Professor Nagamachi conducted research on large vehicles such as aircraft and tankers, and where he also drove the latter vehicle and found that the size of the tire was higher than his height, with eight tires on one side. When he drove a big truck with a 12-gear shift, and changed to the top gear, he arrived at the end of the run, and almost caused an accident.

We were developing a shovel car or excavator using the Kansei product. When we investigated the situation, we found a shortage of drivers and asked their feelings toward the design. Being friendly with them and conducting a survey contributed to the sale.

First, surprisingly, when the actual situation survey was conducted, upon entering the excavator, the drivers take off their shoes and change to slippers. In other words, the driver seat is more like a home for them, which they keep beautiful and clean, and in some cases, it is ideal to work surrounded by music.

From this kind of actual situation survey we started to select the Kansei study words, and again, the needs of the driver became Kansei. The Nakata Engineering Department chief engineer and Iwata design manager participated and have continued to expand the development in this group ever since.

When talking about the driver's Kansei, from the city people point of view, it is not insignificant work but important work, and we want the design to reflect that sense of friendship/relationship. Also, it is hoped that the new design is felt to be smart and have a high-quality appearance. Talking about the construction from the old days, there was an image of mining the land, or classically, the strong form of an angular body of the car in soil color. Professor Nagamachi started with how to break this image.

By the way, Professor Nagamachi was challenged by how to design the angulated body of a car for a sense of relationship and a high-quality form. He also finished the roundness research on how to transform angular things to round. For example, in Figure 5.9, looking at the upper and lower three-dimensional parts, which do you think has an easier feeling?

From the research result, the roundness is about 1 over 12 of the 3D angular part and feels great (sophisticated, friendly), and it comes to a hemisphere shape; when the end of the roundness is increased more and more, one finds a feeling of getting bored.

This principle has also been utilized in cosmetic products of Shiseido, giving the corner of the box a sense of quality and refinement and becoming a little round in design. A box-shaped enclosure of the driver seat in an

FIGURE 5.9
3D Earle Kansei.

excavator is designed in a small format and colored with purple as an idea to express high class and a sense of quality color.

This kind of Kansei and design-related data were provided to the excavator development group, and Avanse Series 1 (Avanse PC45) is shown in Figure 5.10. It received very high marks as an innovative design in the world of construction at that time and received inquiries from a number of vendors.

Because of this, Komatsu was given the Good Design Award in 1992. Then this curved design was applied to a large excavator (PC200t), shown in Figure 5.11. Komatsu also was given the Good Design Award in 1993 again. In addition to the reputation of this large vehicle, aiming to improve the operating rate of the vehicle, the vehicle is equipped with a GPS that can check its movement, but this was an unexpected result.

Only recently, with the construction boom in China, has it been in great demand because some people steal construction assets. Since all Komatsu construction equipment is equipped with a GPS, a stolen vehicle will be found immediately. Komatsu became the first company in the world to install GPS on its construction equipment, giving it a high reputation all over the world. A proposal to mount the GPS is a measure that meets the needs of customers, and the DNA of Kansei engineering has already been implemented by Komatsu.

FIGURE 5.10
Komatsu hydraulic excavators. Avanse series PC45.

FIGURE 5.11
Komatsu excavators. Avanse series PC200.6. Good Design Award in 1993.

5.7 Handrail Development Based on Customers' View

According to the United Nations, when the population ratio of those 65 years old or older exceeds 7%, it is called an aging society; when it exceeds 14%, it is called an aged society; and when it exceeds 21%, it is called a super-aged society. In 1970, the aging population in Japan was 7.1%. In August 2013, the Ministry of Internal Affairs and Communications announced that the population ratio of those 65 years old or older is 24.4%, so we are now in the super-aged society.

In places where senior citizens live, in order to keep a safe and comfortable living environment, barrier-free activity is being implemented. The biggest implementation is handrail installation. At Onomichi City University, associates of an Ishihara assistant professor conducted a survey of those 60 years of age and older in Onomichi City, asking: "Do you use handrails?" Those 65 years old and older who utilize a handrail are shown in Table 5.1. For those 75 years old and above, the usage of a handrail increased. So the handrail is important to them.

TABLE 5.1

Handrail Usage Ratio for Senior Citizens above 60 Years Old

Item	Age of 60–64	Age of 65–69	Age of 70–74	Age of 75–79	Age Above 80	Total
Not used	93.9	81.9	78.9	56.4	40.5	72.7
Used	5.7	16.8	20.3	40.6	42.4	23.2
Requiring long-term care	0	0.3	0.4	1.2	5.7	1.5
No response	0	1.0	0.4	1.8	11.4	2.7

Note: All values in %.

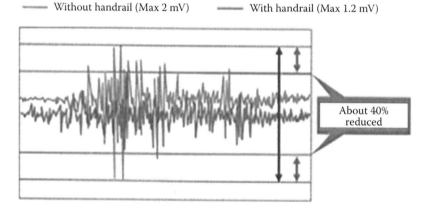

■ Comparison of the load on the lower limbs (power saving) while stepping down from the stairs

※ When the swinging width is smaller, the load becomes lighter

—— Without handrail (Max 2 mV)   —— With handrail (Max 1.2 mV)

About 40% reduced

FIGURE 5.12
Forty percent say that it is easier to get down when holding the handrail, rather than not holding it. (Experiment by Hiroshima Prefecture Health and Medical University, Shiokawa Laboratory.)

An experiment was conducted to understand the effectiveness of the handrail presence. Seventy-year-old senior citizens were requested to go down the stairs using two methods: with and without holding the handrail. As shown in Figure 5.12, a 40% electromyogram (EMG) level of the feet can be reduced by holding the handrail. When holding the handrail while walking down the stairs, it is safer, makes walking more comfortable, and requires less energy. The effectiveness of handrail usage is great.

Per the recommendation of the old Ministry of Construction, a 40-mm diameter of the handrail is desirable. But according to the joint study between Matsushita Electric Works (now Panasonic Electric Works) and Professor Nagamachi, it was found that older women were unable to grip the thick handrail. A few handrail samples were prepared with diameters from 30-mm up to 40-mm, with 2.5-mm gaps, with a total of six types of proper thickness handrail to be held by 86-year-old senior citizens. The investigation results are shown in Figure 5.13. Looking at this, it would be good if the diameter were approximately 35-mm. In other words, the palm size for older women is relatively smaller and reduction of the handrail diameter is necessary. However, for men with a weight of more than 65 kg, the 35-mm diameter handrail is dangerous and can more easily malfunction when held, causing them to fall down. As a countermeasure, Matsushita Electric Works has succeeded in increasing handrail strength by combining three pieces of plywood for the 35-mm diameter.

In addition, the handrails should be mounted close to the wall regardless of rules from the Ministry of Construction. As indicated in Figure 5.14, safe attachment and an easy-to-grip ergonomic location should be in place when considering narrow Japanese-style stairs.

It is also a challenge to mount the handrail height in a house. It is difficult to get one height that functions for both taller and shorter people. It was decided to measure the optimum handrail height by collecting information from 86 senior citizens. A cane-like stick with a spring at the end of the cane with a scale of 1-cm increments was used, and each of the elderly decided the most comfortable height of the handrail. Because the cane is light and spring-loaded, it will determine the optimum height by the elderly person holding it lightly. The result is shown in Table 5.2. A regression equation has been used for these data. Based on the regression equation, each Y can be decided, and the optimum height of the handrail is obtained.

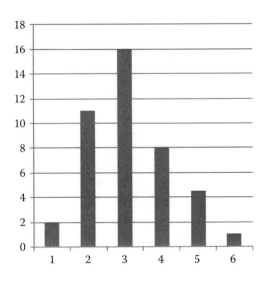

FIGURE 5.13
Sample of acceptable handrail thickness (average 35.4 mm).

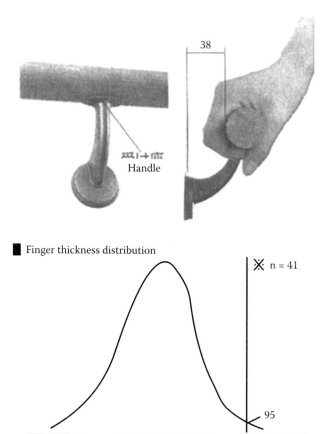

This is the calculation formula from human engineering:
95 percentile × 1.5 (safety factor) = about 38-mm.
Description of the formula: in 100 persons, the 95th person's
thickness value obtained by multiplying the (safety factor) and
coefficient in human engineering.

FIGURE 5.14
Handrail-mounted method using a human engineering device.

When looking at this regression model, the handrail is not decided based on the person's height alone; the height of the elbows is moderately considered. Then, it is not only a horizontal handrail, but also a vertical handrail. It is needed especially when standing up for a stile (Figure 5.15). An experiment was conducted for mounting the vertical handrail position. Mark 0 indicates the position where the testers were sitting at the upper stile. Then we set the handrail forward every 10-cm from the last position. From a sitting position at the stile, we extended the tester's hand while gripping

FIGURE 5.15
Experiment of vertical handrail.

each handrail until he or she was in a standing position. In each scenario, we measured the EMG of the rising action. Details were measured for the supra-spinal muscle of the shoulder, the root flexor muscle of the arm, and the femur biceps muscle of the leg. The result is shown in Figure 5.16.

As indicated in Figure 5.15, to stand up at a vertical handrail from a sitting position, strong physical arm strength is required. In other words, while grabbing the handrail, the body weight needs to be lifted up. However, it is easy to stand up from one's origin to 30-cm handrails and use half of one's energy in the 0-cm position. The reason for this is because the same force toward the hand is being used during holding and raising up the vertical handrail for the 30-cm position (Figure 5.16). Of course, when in the 60-cm position of the vertical handrail, a large force to stand up is needed. It is regrettable that the architect doesn't consider these ergonomic principles. Rather than attaching the same position as he or she does with the vertical stile handrail, it is easy for a senior citizen to use 20–30 cm position toward the front of the handrail.

However, a mounting method that is friendly for senior citizens includes a 40° slant from the stile. When this slanted handrail is used, senior citizens are able to stand up easier due to usage of force at both the hands and feet.

Although the handrail was developed in collaboration with Matsushita Electric Works, Ltd. (now Panasonic Electric Works) and researchers with

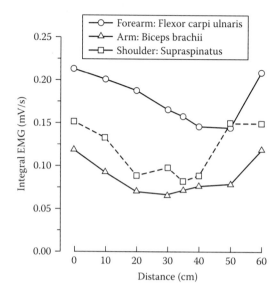

FIGURE 5.16
Muscle potential of the rising activities of a vertical handrail. With a handrail 30 cm away from the stile, it is easier to stand up.

joint development by Hiroshima University, with a total of six group members, this product was the top moneymaker for the company.

Matsushita Electric Works incorporated the concept of universal design and Kansei engineering very early and has been pursuing a variety of new development products. The main focus is the Yokoyama Engineering Department councilor. Tile roofs, gutters, walls (siding), outer structures, baths, toilets, kitchens, storage, lighting, floor heating, and others have used the Kansei engineering philosophy in most products. The companies have thought of a variety of new products using the Kansei engineering concept that do not yet exist in the world.

5.8 Fun Use of a Toilet Foot Step: A Toilet Design

Introducing a new product developed by Matsushita Electric Works that can be used by common people and senior citizens, Nagamachi received a new toilet design project from the toiletries division. There was consultation between the methodology of Kansei engineering and the Yokoyama councilor. This time, it was related to measuring the human buttocks. It is not permitted to measure by directly touching the student's buttocks. The sensitivity measurement method proposed was to provide seven units of ready-made toilets from various makers. Then let the students test by sitting on each unit and evaluating the seating condition according to sensibility words. This method was similar to the experience of the Wacoal project, where one part was related to evaluating the sensibility word and the other part measured the physical measurement of the toilet. The analyzed data would be used as basic data for new products. However, the work was not finished yet. The interpretation of the selected data is important to inspire the desired design (Figure 5.17).

The human engineering laboratory of Hiroshima International University set up a sample of seven toilet units that were purchased from a variety of makers. Eighteen male and female students took part as testers. Forty Kansei words were selected for use as the Kansei word related to the comfort of the usage of the toilets. They are as shown in Table 5.2.

Each tester was given the evaluation sheet and sat on the seven units randomly, according to the table in the evaluation sheet, and evaluated the feeling of sitting on each unit. A statistical analysis was done by collecting the evaluations of all voters. First, analysis was conducted of the variance, a Kansei word of what is good in every toilet. Then, by using the same data, the factor analysis was measured and then the Kansei factor of the

FIGURE 5.17
Toilet seat scenario experiment.

TABLE 5.2

Toilet Evaluation Sheet

Name			Date	/ / /
Age		Sex Male, Female,		Sample No.( )
Height	cm,	Weight		kg
1. Easy to sit	☐ ☐ ☐ ☐ ☐			Not easy to sit
2. Large	☐ ☐ ☐ ☐ ☐			Small
3. Wide	☐ ☐ ☐ ☐ ☐			Narrow
4. Easy	☐ ☐ ☐ ☐ ☐			Not easy
5. Balanced	☐ ☐ ☐ ☐ ☐			Not balanced
6. Safe	☐ ☐ ☐ ☐ ☐			Unsafe
7. Not tired	☐ ☐ ☐ ☐ ☐			Tired
8. Soft touch	☐ ☐ ☐ ☐ ☐			Hard
9. Steady	☐ ☐ ☐ ☐ ☐			Not steady
10. Fit	☐ ☐ ☐ ☐ ☐			Not fit
11. Comfort	☐ ☐ ☐ ☐ ☐			Discomfort
12. Easy to urine	☐ ☐ ☐ ☐ ☐			Not easy
13. Easy to stand	☐ ☐ ☐ ☐ ☐			Not easy
14. Want to use	☐ ☐ ☐ ☐ ☐			Don't want

word interpreted. On the other hand, the physical measurements of the seven toilet units was measured and a list created. The measurement of the physical measurements in this case is 24 types; part of them are shown in Figure 5.18.

After that, as the final statistical analysis, the most important Kansei words selected by the designer (Kansei factor) and the words that have the strongest relation to physical measurements were identified through statistical analysis. The results of the data, together with the designer, were translated into the design form. In terms of Professor Nagamachi's experience with new product development so far, most of products are being led and decided by the new design. Based on the design blueprint, the technical person from Matsushita Electric Works will place them in the Kansei Human Engineering Laboratory. The final mock-up was completed based on the analysis of seven ordinary toilets that were adopted for factories.

After a new sample is completed, verification needs to be conducted to validate the suitable level of comfort, which is what Kansei really aims for. A survey is conducted as a verification experiment by the addition of changes in the seven units. The first validation test of an ergonomic surface was added on top of the sensitivity evaluation (refer to Figures 5.19 to 5.21).

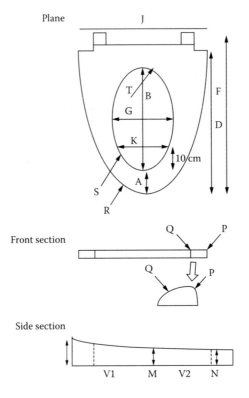

FIGURE 5.18
Part of a toilet's physical measurements.

FIGURE 5.19
Overall picture of the new toilet.

FIGURE 5.20
Result of raising the top cover of the new toilet.

FIGURE 5.21
Side view of the new toilet.

First, as shown by a Kansei evaluation test, the result of analysis of variance shows that the new toilet items all scored best in sensibility. The first ergonomic analysis was to validate that the testers should be comfortable to sit. A body pressure measurement device was used to examine body pressure dispersion, which is called force sensitive application (FSA), and checked the pressure of the body on the seat of the toilet.

Put an FSA on top of the seat surface and ask the tester to sit on it. Then, by resting the human body on top of the computer, where the bone in the buttocks hits strongly will appear. In other words, it will illustrate the pressure of the body that receives the most pressure.

This is shown in Figure 5.22. A body pressure variance value of seven units used in the experiment by other toilet makers shows results similar to those of the right side of Figure 5.22. The black line is prolonged to the thigh of the leg from the waist muscle; this is because when sitting, the direction of blood flow is obstructed toward the foot. On the other hand, the left part of the figure shows the seat-sided pressure variance value of the new toilet, and the black short lines come out only from the waist bone muscles to the leg. In other words, the new toilet shows that the sciatic nodule well supports the body and the blood is flowing in the direction of the foot. This achieved an emotion of "sitting feeling good." Everyone who sat on the new toilet would say, "I wanted to read a book while sitting on this."

The second ergonomic validation point is whether the new toilet has become a friendly product for the elderly. In fact, before the product development of the new toilet, Matsushita Electric Works had mounted armrests for senior citizens on the toilet. Their objective is to allow senior citizens to stand up more easily by gripping the armrests. It became a little easier, but there were still complaints.

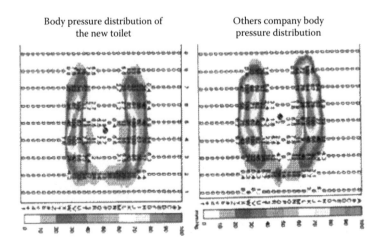

FIGURE 5.22
Comparison of body pressure variance.

To further improve the product, the seat was tilted forward in the direction of the bearing surface to feel comfortable. Senior citizens between 70 and 80 years old took part in the experiment. Quadriceps muscles, the gastrocnemius muscle, and the soleus muscle were connected with an electrode that measured the EMG while standing up. The result was incredible. The EMG during stand-up from the new toilet was reduced to about 1/10 compared to the horizontal seating surface of toilets from other companies. Also, one of the senior citizens agreed to say that it is very easy to stand up.

By applying ergonomics and Kansei engineering, a new product, such as the example, can be developed. This new product was named Torres (TRES). The TRES sold very quickly for the manufacturing company and its production became very busy because of the customers' rush orders.

5.9 Decide the Brand Name by Kansei

There are a variety of ways to determine the name for newly developed products. It is very difficult to predict which method can lead to the most profitable result. The following research shows that when the brand name is read, it should sound as though it expresses the characteristics and content of the product. For example, a parent is considering what name to give his or her newborn child. In most cases, a name is given based upon wishing the baby an ideal way of life. Kanji character is also considered. When the name is spoken, it's a great success if the name has an ideal impact on way of life of the baby. The same goes for new products. It is important to give customers an image of the product when they hear the name of the product. As an example, the Wacoal Good-Up Bra can give the image of position shape up. Tuning of a new product and branding using Kansei will definitely increase sales.

When breathing, the air goes through the trachea and vibrates the vocal cord. The breath can be sent to the mouth and nose and change to sound. Even when the mouth is narrowed or widened, the voice is changed, as well as when the mouth is opened widely or closed. With loud laugh sounds (ahaha), the mouth opens widely, and with quieter laugh sounds (ihihi), the upper and lower teeth are aligned. When producing voice to match with Kansei, the breathing method is controlled by the mouth palate pattern (refer to Figure 5.23).

The next study shows the relationship between tone and voice. The relationship between the voice and mouth shape can be displayed as shown in phonetics in Table 5.3. Breath goes through the nasal cavity when one says MA-MI-MU-ME-MO or NG, and the voice becomes different when releasing and biting the teeth. Table 5.4 shows the relationship between the voice and palate shape. This is called the *point of articulation*. With breathing through the nose, stopping the breath and then releasing it at once will give

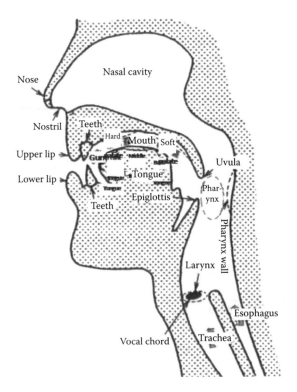

FIGURE 5.23
Anatomical drawing of the palate.

TABLE 5.3

Phonetic Symbol Table

		Bilabial	Alveolar	Palatal Alveolar	Velar	Velar Uvular	Glottal
Consonant	Nasal	M		n		N	
	Occlusive	p, b	t, d		ç	k, g	
	Fricative		s, z	ʃʒ	c		h
	Approximant				j	ɰ	
	Affricate		ts, dz	tʃ dʒ			
	Flow voice		ʃ				

various sounds, as shown in Table 5.5. Although it is quite complex to put out a shape and breath of the mouth, phonetics was organized as in these tables. Based on the data, the relationship between the type of voice and the shape of the palate is converted using artificial intelligence rules.

Next, there will be very complex analytical work. One of the monthly magazines of Bungei Shunju was selected, and all the words related to Kansei

TABLE 5.4

Classification by Articulation Types

Name		Speaking Method
Bilabial	Pa Ha	Open both lips after closure
	Ma	While both lips are closed, pass through the breath to the nasal cavity
	Fu	With a small gap between both lips, generate breath friction
Alveolar	Ta Da	Release after closure in the breath and gum tongue tips
	Ra	Pass through to the nasal cavity after closure in the breath and gum tongue tips
	Na	Pass through breath in between the gum and tongue gaps and generate breath friction
	Sa Za	
	Tsu Tzu	
Palatal alveolar	Shi Ji	Put the tongue tip in between the gum and hard palate and generate breath friction
Palatal	Ki	Release breath after closing both sides of the tongue and hard palate
	Hi	Narrow the front tongue surface and hard palate, and generate breath friction
	Ya	Narrow the front tongue surface and hard palate to the extent that breath friction is unable to be generated
Velar	Ka Ga	Release after closure of soft tongue and inner tongue surface
	Wa	Narrow the base of the tongue and soft palate, and generate the breath friction
Uvula nasal sound	Ng	Back tongue surface close to the soft tongue at the back of the tongue surface (inner tongue surface), and pass through the nasal cavity
Glottal	Ha He Ho	Narrow the glottis without vibration of the vocal cords and generate friction

words were jotted down from the first page until the end. From hundreds of words, *like* and *dislike* were excluded and only the words suitable to human emotional expression were kept. Then, they were transformed to Kansei evaluation data, and using factor analysis, we picked the words according to some factors and reduced the list to 68 terms.

Each of the 68 words, as mentioned above, is put into a classification table and then a list is created of articulation point and articulation type. This is the basic database for the sound of words. Back to the previous factor analysis, detailed study of each factor and sound evaluation of the words was performed and 20 positive vs. 40 negative words were selected that are appropriate Kansei words. This is the basic measurement scale for evaluating the brand in the future. The list is shown in Table 5.6. Next, to judge the impact to a customer's ear of a person's name or product branding, evaluation with a 40-item Kansei word scale was used with the plus and minus side method.

TABLE 5.5

Classification by Articulation Types

Articulation Types		
Name		**Speaking Method**
Nasal	Ma Na En	Breathe using a place in the oral cavity or head to close it and flow out using the nasal cavity
Occlusive	Pa Ba Ta Da Ka Ga	Breathe using a place in the oral cavity or head to stop a while and one time immediately breathe again
Affricate	Tsu, Tzu, Chi Ji	Stop the breathing for a while, and start to open it slowly
Fricative	Sa Za Ha	Breathe using a narrow path to open its surroundings
Approximant	Ya Wa	Interfere until the breathing almost does not have friction
Flow sound	Ra	Narrow the oral cavity until the breathing almost does not have a frictional degree

TABLE 5.6

Influence of Kansei Evaluation Scale (40 Words)

1	Soft	——	Hard	21	Elegant	——	Not elegant
2	Bright	——	Dark	22	Modern	——	Not modern
3	Spread	——	No spread	23	Cute	——	Not cute
4	Unique	——	Not unique	24	Simplicity	——	Not simple
5	Open-minded	——	Not open-minded	25	Japanese	——	Foreign
6	Heavy	——	Not heavy	26	Attractive	——	Not attractive
7	Refreshing	——	Not refreshing	27	Romantic	——	Not romantic
8	Clear	——	Ambiguous	28	Good feeling	——	Bad feeling
9	Simple	——	Complicate	29	Dry	——	Wet
10	Gorgeous	——	Plain	30	Adult	——	Young
11	Cold	——	Warm	31	Impressive	——	Not impressive
12	Personal	——	Not personal	32	Beautiful	——	Not beautiful
13	Sense of dance	——	No sense of dance	33	Intellectual	——	Not intellectual
14	Good influence	——	Bad influence	34	Healthful	——	Not healthful
15	Aerodynamic roundish	——	Not roundish	35	Smooth	——	Thin and bony
16	Friendly	——	Unfriendly	36	Monotonous	——	Not monotonous
17	Masculine	——	Feminine	37	Fresh	——	Not fresh
18	Sense of flow	——	No sense of flow	38	Lively	——	Quiet
19	Sharp	——	Not sharp	39	Calm	——	Restless
20	Strong	——		40	Luxury	——	Not luxury

Finally, an artificial intelligence diagnostic system is developed to find the human feeling when hearing any brand name. First, create a system made up of four characters of meaningless words. Then let the tester use Table 5.3 to evaluate the Kansei. Use the result and compose an eight-character word. Build this evaluation to construct artificial intelligence (hierarchical fuzzy measure and integral model). Due to a complex process (omitted here), a three-layer model was developed, as shown in Figure 5.24.[3,4] A Matsubara assistant (currently a Hiroshima City University professor) and Maeda graduate student built the model patiently.[1-3]

Next is the result of the diagnosis hierarchical fuzzy measure and integral model (HFI). First, to indicate how well the model is diagnosed, all readers' understanding of *jigoku* ("hell") was used and the impact of the sound presented. The result is shown in Figure 5.25.

This computer system doesn't know what *jigoku* is, but most of the emotional words point to the negative, which is on the right side, such as dark, closed, cold and sounding bad, indecent, and ugly—all considered negative images. There is no result here, but *paradise* has been used for evaluation as the opposite word for *hell*. The system surprisingly analyzed like a human sense in its way of thinking.[5]

Next, let's analyze a person's name. The name of a currently young, popular, and famous model, Suzuki Nana, is being selected in Kansei. What should be noted here is that it is not related to the family, but it is the last name and it cannot be applied to the analysis. The given name by the parent is subject for evaluation. So, how about *Nana*?

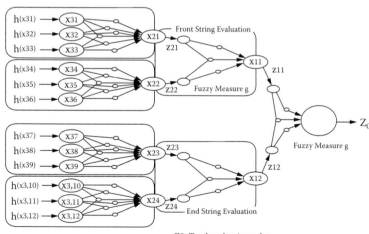

Z0: Total evaluation value
Zij: Each part evaluation value
Xij: Each part evaluation item
h(xij): Evaluation value for each part evaluation item (data input)

FIGURE 5.24
HFI model diagram for word impact evaluation.

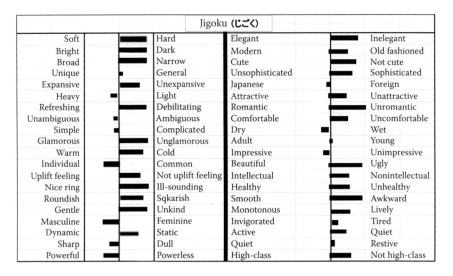

FIGURE 5.25
Kansei evaluation of Jigoku.

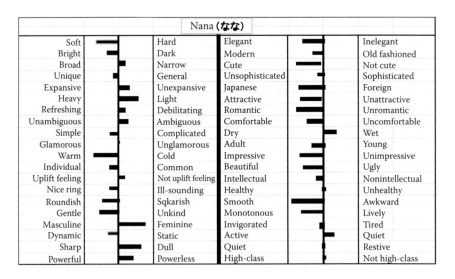

FIGURE 5.26
Kansei evaluation of *Nana*.

Analysis results, as shown in Figure 5.26, consist of soft and warm and roundness, gentle like a girl, cute and elegant, and solid compliment. It becomes the Suzuki "Nana's" Kansei evaluation. The number 1 of the top 10 best names for girls by the Meiji Yasuda Life Survey for 2012 is *Yui*. The Kansei evaluation of this name is not as good and came out as a calm Japanese girl, but there is no other remarkable specialty.

Also among the top 10 best names is *Ren*; although this analysis is for a modern individual, the result indicated it is somewhat feminine and not a distinguishing feature. It shows that it may be good to avoid giving a name like this.

In the United States, a very popular girl's name is *Lily*. If we input the name *Lily* into the brand name system, it is able to diagnose the pronounced image, although it is an English name. The result is shown in Figure 5.27. The name *Lily* gives people the sound of bright, lovely, clear, feminine, modern, romantic, young, charming, and smart. People feel it as a good sound.

Let us analyze the brand name that was given to a real product. What is the sound of the small type of the Nissan March? The analysis result is shown in Figure 5.28, and it sounds like a rounded, gentle, warm, feminine, cute, and elegant car. The design of the March is also rounded and feels simple; the design and the brand name match and are popular among the ladies.

A lot of the Kansei artificial intelligence system has been developed, except for a sound diagnostic system of words. There are a handle Kansei system for Nissan Motor, a design system for automotive interiors for Isuzu, a kitchen design system for Matsushita Electric Works (ViVA), a cockpit design system for Komatsu, a color coordination system for Sharp, and others. With a combination of Kansei and artificial intelligence, there are a lot of interesting mechanisms occurring that are not only pleasing to designers but also effective tools in business. The kitchen system of Matsushita Electric Works is very successful, and a lot of visitors from domestic and external customers visit it every day.

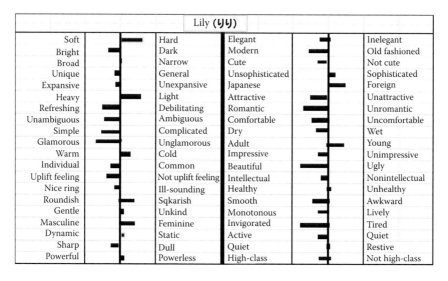

FIGURE 5.27
Kansei evaluation of *Lily*.

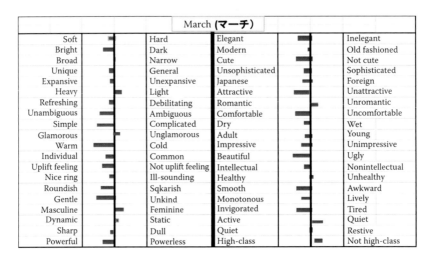

FIGURE 5.28
Kansei evaluation of *March*.

5.10 Confectionery Package Design at Nestle

In February 2007, Nagamachi made a presentation on Kansei engineering at the PACE Conference Packaging Design International Seminar in Paris. At this event, he met Nestlé packaging and design director Anne Roulin, and Daniel Magnin, head of packaging and design for the Nestlé Confectionery R&D. Both of them were very interested in Kansei engineering and wanted to know more about Kansei engineering and its methodology. There was a good discussion at the venue. Later, Mr. Magnin visited the Hiroshima office and a more in-depth discussion resulted; as a result, a formal contract was signed to introduce Kansei engineering to Nestlé Confectionery Business in March.

For phase 1 of this project, five to six lectures were given to the packaging and design department, marketing department, and their administrators, explaining Kansei engineering in detail. They were new to this topic, and lots of questions were raised; the most challenging questions and discussions were from the marketing group. Marketing research is to conduct surveys in order to know the consumers' buying behavior, which makes is quite different from Kansei at its roots; Kansei is to investigate images that trigger the same emotion shared by the consumers, and then convert these images into design data. Eventually, they understood after several meetings. Then it was time to conduct lectures for the designers: Ben Mortimer (Nestlé Confectionery R&D) and Pascual Wawoe (freelance). They are passionate and insightful persons, and we enjoyed every minute. In conclusion, they decided to upgrade the package design of After Eight®, a chocolate with high evaluation from Nestlé, by using Kansei engineering.

5.10.1 Step 1: Gathering Research Candidates

First, we had to determine the more well-known manufacturers and their chocolates in Europe. Keeping those brands in mind, we bought more than 100 types of chocolate from various high-end supermarkets. We sorted them accordingly: items rival to After Eight, eye-catching packages of shape, size, and color (Figure 5.29).

Even after the sorting, we still had too many items; we needed to select only the best brand, shape, size, color, and package design from those groups. We selected 21 items, including several types of After Eight, in this round (Figure 5.30).

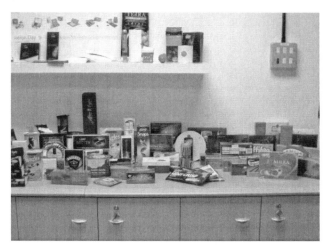

FIGURE 5.29
Chocolate package samples in the first stage.

FIGURE 5.30
Test samples for packaging design survey.

5.10.2 Step 2: Select Kansei Words

In Kansei engineering, we conduct our research by evaluating Kansei words, words often used by the target consumers (adjectives or descriptions or key words). Therefore, selecting the right key words is very important. Two young designers and Nagamachi selected many key words related to package design for evaluation.

We prepared a questionnaire with 23 questions with 7-point scales with the final selection of key words, and they were divided into three stages. The first stage of evaluation was "imagine the chocolate inside the box by observing the package design." The second stage was to evaluate the package design related to the key words. The third stage was the "emotion of opening the package and seeing the chocolate inside or on the package itself."

At this point, the package samples and the questionnaire were ready. Nestlé chocolate sells best in the United Kingdom and Germany; whether separate package designs or one unified design should be used became another issue. It was decided that a decision would be made after conducting research and analysis in both countries. With the help of native speakers, the questionnaire was prepared in both English and German.

5.10.3 Step 3: Conduct Evaluation Research

Recruitment was done through both countries' newspapers without identifying Nestlé; together, 200 participants were recruited, and samples were given to them randomly during the evaluation (Figure 5.31).

FIGURE 5.31
A scene of Kansei evaluation.

5.10.4 Survey Data Analysis

In the Nestlé chocolate package design analysis the following was done:

1. Analysis of the relationship between Kansei words: This process was to confirm the emotional similarity between the 23 Kansei words used. There might be very high similarity between the Kansei words, even in the UK and Germany.

2. Factor analysis of both countries' data: With the 23 Kansei words, a few special clusters were found, and it was discovered that these clusters (structural factors) were very similar. As a result, we were able to consolidate the two countries' data into one set.

3. Principal component analysis (PCA): This was not a simple PCA. A PCA was used to discover the Kansei words structure (emotion), but then we put a 21-sample axis on top of this structure, so we could see which sample was overlapping with which emotion by searching from the same coordinate axis. This analysis can define the evaluation of each sample; package designers can validate the ideal design factors' location from the consumers' point of view.

   In order to conduct the PCA, we needed to convert the shape, color, size, and package design characteristics of the 21 samples into sample data. We called them the physical traits. They were analyzed on the same coordinate axis.

   From the result, we discovered that some of the applied Nestlé chocolate characteristics were a little bit distant from the attractive position. In other words, we understood that there was room for improving the package design by using Kansei engineering.

4. Partial least squares (PLS) analysis: PLS was used to validate the ideal Kansei words (consumers' emotion) and convert them to physical design factors for implementation, by using least squares methodology, which also draws an applicable conclusion to Kansei engineering.

   Conducting the above-mentioned analysis with sample evaluation data collected from the participants, we created a PLS chart. Within it we could validate the design specs or design elements we must implement in order to achieve an attractive package design. We should be able to create a suggested design that can generate more sales by using this enormous information.

At the Nestlé Product Research Center in York, England, with these enormous design factors, Nagamachi worked with two designers; we each started to draw our new design. Then we put tens of sketches on the wall, observed them from a distance, and finally created a new packaging design. In order achieve this, the three of us had numerous discussions, optimized our design ability, and created an ideal design that could trigger consumers' curiosity.

5.10.5 New Package Design Presentation

In mid-November 2007, Nagamachi went to London to present the final result of the Kansei engineering research. We also prepared a mock-up sample to show the audience at the end of the presentation. The reason for having this presentation in London was that all our audiences included designers, marketing staff, and the manager who supervises the design department, so it was convenient to meet there. Ben and Nagamachi spent over an hour explaining the activities conducted and our consolidated analysis result. We were really concerned about the response of the audience. We explained the research in detail, and we sensed a negative response from them, so Ben and Nagamachi looked at each other and then revealed the mock-up sample and said, "This is the result of the research."

The floor became silent for a moment, and then Nagamachi stole a glance at the audience. All of them stood up and gave us a great round of applause. We received a standing ovation.

Mr. Magnin, Ben, and Nagamachi were so glad to know that they all understood. Figure 5.32 shows a picture of the mock-up.

Later on, Ben and his team revised the design and created two more After Eight packaging designs, as shown in Figures 5.33 and 5.34, which we often find in department stores or international airport stores.

A step-up design from our mock-up (Figure 5.32), an individual pack design with a smaller portion of chocolate can now be found in international airports. Nagamachi knew this design was also another step up from the basis of Kansei engineering research.

For Nestlé, based on this successful experience, they recommend using Kansei engineering for new product development. They even developed a Kansei manual for implementation and developed more new products.

FIGURE 5.32
After Eight packaging design using Kansei engineering.

FIGURE 5.33
New packaging design of After Eight.

FIGURE 5.34
The recent new design of After Eight.

The following is the comment from Nestlé:

> Nestlé has been exploring the opportunities that the Kansei engineering approach brings to packaging design, with the support of Professor Nagamachi as well as Design Perspectives from Faraday, and has done pioneering work in applying this technique to package design. Today, Kansei is one of the specific tools in Nestlé's Packaging and Design toolbox for consumer centric development. However, it does not replace the need for experienced industrial designers, graphic designers or packaging engineers whose skills are still needed to turn Kansei-generated data into tangible product packaging.

5.11 Curable Bedsores

Japan is a super-aging society. The population ratio of those 65 years or older (the super-aged society) is 24.4%, as announced by the Ministry of Internal Affairs and Communications in August 2013. A major issue due to aging of the bedridden elderly is bedsores. Regardless of age, bedsores (pressure sores) are not limited to the elderly; they are also a big problem for the disabled who use a wheelchair. Due to lying for a long time, bedsores occur and trouble medical personnel.

Pressure sores are the pressure on the skin on the bone under the weight of the body when lying, resulting in a vascular disorder taking place on part of the skin. The failure that occurs, the extended bone, often hits at the sacral area of the buttocks (63%), calcaneus of the heel (7.8%), large trochanter of the hip (6.9%), as well as the occipital region of the head and shoulder blade of the shoulder (Figure 5.35).

As defined by the Japan Society of Bedsores, bedsores are caused by pressure applied to the same side, which causes a blood circulation disorder at the compression site that results in necrotic tissue. Factors related to bedsore occurrence include:

1. Continuous local pressure
2. Friction
3. Shear stress
4. Wetting of the local area
5. Malnutrition

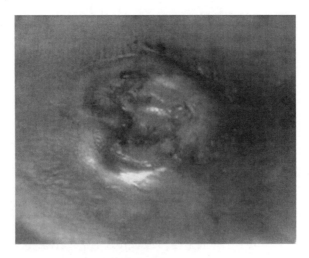

FIGURE 5.35
An example of a bedsore.

Factors 1 to 4 are especially the result of the physical action of external handling of the material between the mattress and the body.

Assuming that bedsores are a medical problem, at the same time it can be said that between them and the human being there is a human engineering problem. Here is where we can treat and help prevent bedsores. Ergonomics is studied to produce the desired relationship between an object and human (mind and body). We build a system and develop tools for that. Therefore, bedsore measures are a topic of ergonomics, as well as a medical problem. The Japan Society of Bedsores presents a benchmark index where the average body pressure distribution value on a mattress should be less than or equal to 32 mmHg.

The mattresses that are used in hospitals (except the air mattresses) were purchased separately from 12 types of markets, and then the body pressure of six people was measured using a tool to measure the body pressure distribution value of the FSA. Among them, some had slightly higher than the reference indicators, but in most cases, the values were below the standard. However, we discovered a problem point: the concept of decompression among the medical personnel.

In mechanics, decompression is the case when the body is lying on the mattress and the whole body is treated on the mattress; in other words, some areas of mattress not treat by the whole body on the mattress should function to receive the pressure of the body. This means that you wrap the entire mattress around the body. This is an inconvenience from the viewpoint of sleep science. It will vary from person to person, but in one night, people will turn over 20 to 60 times, and this rolling over is very difficult when the mattress envelops the body; the person would feel tired upon waking up in the morning. In other words, from the engineering perspective, we are required to take the weight of the body horizontally on the entire mattress. This is called *body pressure dispersion*. There were no mattresses to fit to this condition in the survey.

So, Professor Nagamachi went into the study of the material that makes up the mattress. Fortunately, he found it right away. It was a polyester material (Breathair) sold and manufactured by Toyobo Polyester. It is shown in Figure 5.36, where the material is synthesized by complicated entangled pipes made of polyester and has a characteristic of high rebound force. It is most suitable for mattresses where some of the pipe is in a one-by-one vacuum condition and the hollow material in the pipe is buried and has flexible characteristics in fulfilling pipe. The characteristics of Breathair are:

1. Its entangled pipe structure has a strong elasticity.
2. It is highly breathable because of its mesh structure.
3. It is both easy to dry and easy to wash.
4. It is easy to carry for a woman because it is lightweight.

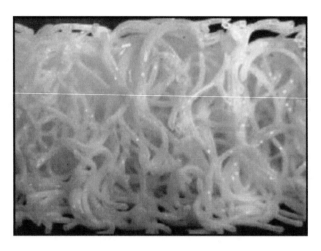

FIGURE 5.36
Breathair material.

It has other characteristics as well. For example, we easily can clean the patient in the shower, even if urinary incontinence is present. However, even if the same cannot be said for Breathair, the search for the optimal Breathair structure begins here.

Breathair is the structured production and involves amounts of polyester, differences in density, thickness of the pipe, filled or hollow, Breathair thickness, and other functions that are completely different.

First, in 2006 the researchers obtained Breathair mattresses of different functions and various types, and then began research on the differences in function. In addition, in 2008, Toyobo and Panasonic Electric Works (formerly Matsushita Electric Works), in a three-way collaboration with the Hiroshima International University Human Kansei Engineering Laboratory, started a comparative study on mattresses and polyurethane material of 12 types of commercially available and 70 types of Breathair mattress material (including two-layer, three-layer, etc.).

The research and analysis is measured using the body pressure dispersion value by FSA:

1. Comfortable is good.
2. Sinking is small.
3. Seems to sleep well.
4. Easy to roll over.
5. Comfort.
6. High quality.
7. Elegant.

A survey of Kansei is conducted using a five-step rating scale providing the previous seven types of Kansei words. The testers are men and women (including the elderly) of 40 to 108 kg body weight. This was very tough work.

The FSA measurement was usually used in this study, as mentioned previously, and the mattress needed to be able to support the total body weight; in other words, it is hoped that when lying on the mattress, it is easy to roll over and feels free and supports without sinking. This type of Kansei evaluation was added. In other words, normally the mattress evaluation was done only in a one-dimensional survey of the FSA, but we tried the world's first two-dimensional study of the sensitivity and FSA measurement.

Using 70 types of Breathair material for the tests, the body pressure dispersion value needed to be close to or lower than 32 mmHg. There is a sign of Toyobo, but Professor Nagamachi used P and Q. P is enriched pipe material having a thickness of 3 cm, and Q has a thickness of 5.5 cm. For single units, the researchers had characteristics of "do not sink when used" and "easy to roll over," but when P was put at the upper level and Q at the lower level and added (8.5 cm), the great FSA measurement value and numerical Kansei were developed.

The FSA value will change based on the characteristics of the fabric of the side place to wrap up the mattress and the final products developed from the three types decided upon. These three new product types and the mattresses of 12 types of polyurethane that were determined by the market resulted in the verification by statistical analysis shown in Figure 5.37,

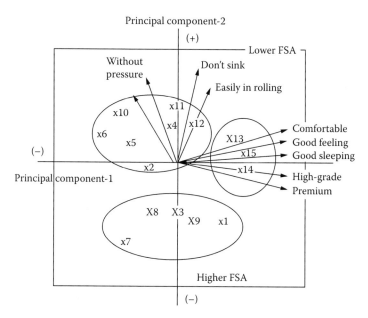

FIGURE 5.37
Principal component analysis of 15 different mattresses.

which is the result of the principal component analysis of the 15 different mattresses including three new types that were selected. Interesting results of the analysis were obtained, as shown in the figure, based on the in-depth measurement of the Kansei and FSA measurement data.

The horizontal axis represents the Kansei evaluation data (principal component 1) axis, and the vertical axis is the average value for the maximum pressure dispersion value of the FSA measurement of main component 2. To make it easy to understand, let's start by discussing two main components. The vertical axis indicates that the body pressure dispersion value of the product that exists in the upper part is low, indicating that the product at the bottom is high. Fifteen types of mattress are numbered along this axis: X4, X5, X6, X10, X11, X12, and so on. They are one group and belong to the class of the lower body pressure dispersion value. We folded into these the desired position as the body pressure dispersion value, and named this cluster 1.

On the other hand, X1, X3, X7, X8, and X9 on the lower side have become lump mattresses of five types; these are the mattresses whose distributed body pressure was undesirable because it was higher than that defined by the Japan Society of Bedsores. It is named cluster 2. With redundancy, Professor Nagamachi began the bed study in 1965, and because it was originally developed for the first time in Japan, the body pressure dispersion measurement instrument in this case became his specialty.

A totally different group of mattresses is distributed on the right of the horizontal axis (principal component 1) in the figure. This is the new product mattress group developed; they are in the desired position for the body pressure dispersion value. This group is different from the other clusters and has been evaluated as "comfortable and easy to roll over and very comfortable." This group is named cluster 3. The characteristics of cluster 3 are intermediate values of moderation, where you can found the desirable characteristics of emotion all together. Clusters 1 and 2 are on the Kansei axis far from the optimal position and represent a product that is not comfortable and difficult to turn over on. Based on this, the new products of P/Q that Professor Nagamachi produced using the Breathair material were verified to be good new products that did not already exist in the world.

The second important validation is the most important. Bedsores indicate that blood flow is inhibited. Permission was asked of an 82-year-old tester with a sacral projection to measure his blood flow using the mattresses that were part of in this experiment. With the help of the Osaka National Institute of Advanced Industrial Science and Technology, the blood flow measurements were realized.

The upper part of Figure 5.38 shows the blood flow data for 30 minutes when the same testers were lying on the commercial mattress, and it shows a situation where the flow value is not substantial. The part on the bottom of the figure shows the new product that has been designed and the testers

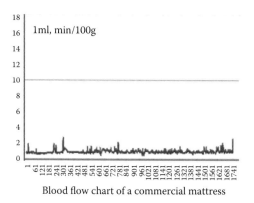

Blood flow chart of a commercial mattress

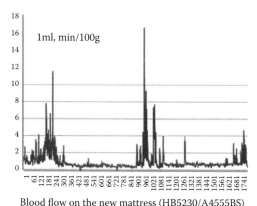

Blood flow on the new mattress (HB5230/A4555BS)

FIGURE 5.38
Blood flow situation of new products and commercial products.

lying for 30 minutes on the two-layer Breathair mattress, and it is understood that the blood is flowing when the tester smoothly rolls over.

The new product mattress is a high-elasticity material, and patients can easily roll over and blood flow occurs each time. As described before, although people roll over 20–60 times during sleep overnight, with the high-repulsion mattress, you can roll over several times and oxygen is carried to the skin blood flow easily. As a result, bedsores occur less frequently. The mattress can also comfortably support a change in body position. For the same reason, the product body pressure distribution is received by the entire mattress with high repulsion, making it possible to comfortably sleep.

The third verification is that, with the help of new products for bedsore patients, on-site, on-the-spot investigation is done to see the changes. We donated ten mattresses each to four locations including the national hospital and a public facility. Each facility was led by the plastic surgeon or deputy hospital director; a doctor, bedsore-certified nurses and other nurses, a pharmacist, and a physiotherapist with a team made of, for example, occupational therapists under the leader and on the condition that the care follow the protocols of the Japan Society of Bedsores—treatment between 6 and 8 months. Meanwhile, each medical leader had a note to capture the progress of bedsores with a camera.

Before this investigation began, approval by the ethics committee was received at each facility, as well as the consent in relation to the personal information of patients. The daily therapy monitoring took the data in accordance with the provisions of the Design-R over the Japan Society of Bedsores (Design-R) and created a graph of the variations. The use of statistical analysis was planned by gathering patient data of all facilities since the progress of each individual is different.

The Design-R is defined by the Japan Society of Bedsores (bedsore course evaluation table) based on the depth (the depth of the bedsore), exudate, size, inflammation, granulation, necrotic tissue, and depth, and these seven indicators are to be recorded. When arranging the first letter of each, it becomes Design-R.

It was found that taking the statistics of all patients is difficult, and data of Design-R from the national hospital bedsore patient population is shown in Figure 5.39. Although at the start of the survey the bedsore patients consisted of 15 people, deaths or change of hospital reduced that number to 7. Upon using the new products, the bedsores of five people, as early as 1 week and by 5 weeks at the latest, were completely cured (short-term cured group). Two patients experienced long-term cures; one of them was cured in 4 months, and the other person was cured but had a recurrence, finally being cured at about 5 months. After this, of all the bedsore patients that were brought in, all of them were cured by August 2013, and the number of bedsore patients

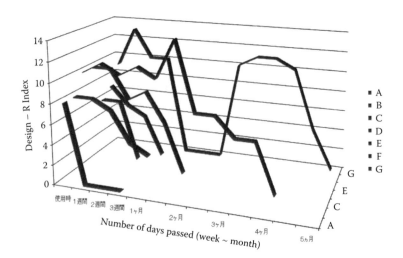

FIGURE 5.39
Design-R values of seven patients.

dropped to zero. The same thing happened at four other locations until bedsore patients no longer occurred in hospitals.[6,7]

Bedsores are an extremely painful condition, but they are not impossible to cure. Bedsores can be solved medically, and ergonomics but Kansei engineering points of view are still insufficient. Joint research with the Ministry of Health, Labor and Welfare was requested but grants were rejected, as bedsore treatment was claimed to be difficult. The officer was also required, as medical personnel, to learn this new principle by all means. Indeed, it has developed a reputation very quickly overseas. It should be noted that the mattress is produced only in HappiOgawa and Panasonic Electric Works, as well. Other mattresses are independent of this principle.

5.12 Interior Design of Boeing 787

One day in 2003, Professor Nagamachi received a phone call from his American friend living in Michigan, Glenn Mazur. He said, "Boeing would like to use Kansei engineering for aircraft interior design, but do give some guidance." Professor Nagamachi had experience on the interior design of a car or house; this was interesting because it was his first time doing the interior design of an aircraft. The offer was accepted.

Glenn Mazur is a professor at the University of Michigan and also the executive director of the Quality Function Deployment (QFD) for the Secretary General of the International Conference and was a student of a Japanese language course when Professor Nagamachi was a researcher at the University of Michigan. Glenn started learning Kansei engineering in response to Professor Nagamachi's invitation to the QFD International Conference in 2004 on conducting a Kansei engineering workshop, in which Glenn also participated.[8] After a while Professor Nagamachi received a phone call saying, "I will be the instructor. Please guide me." After more time passed, Professor Nagamachi received another phone call from Glenn, with him saying, "I don't know what to do. Please write the method on paper and also send it to me by email." Professor Nagamachi felt annoyed, but Glenn is his friend and junior, so Professor Nagamachi described the method in detail and sent it to him.

Professor Nagamachi did not describe the methodology of researching the interior of a car in this document. The easiest way to do that is to have 10 units of various cars, arranged with the help of a rental car company, and ask 50 testers to note Kansei words (about 30 words) while they are sitting inside the car, and let them answer questions related to the interior of the car. Then all the surveys are collected and statistical analysis performed with Kansei engineering to see what kind of design was seen from a Kansei perspective. But for Boeing and other aircraft manufacturing companies, it is impossible to arrange to have 10 units of aircraft. So Professor Nagamachi thought of an idea based on the following steps:

1. To explain Kansei engineering in a simple way, start with the words chosen to express the Kansei from frequent flyers who use the aircraft. This is called Kansei words.
2. Provide several sample designs and transform the data to the physical characteristics.
3. Evaluate the sample designs by the frequent flyers using the Kansei words prepared.
4. Determine by statistical analysis the relationship between the physical characteristics and the design Kansei words (multivariant analysis in most cases).
5. Write down a new design with reference to the physical characteristics most relevant to the words of sensibility that we want to achieve.

First, Glenn collected a few dozen customers that often use aircraft and conducted a survey of their emotions (Kansei) when they are on the plane. When the result was collected, Kansei words for investigation were determined. Next, questions were asked of the customers that related to the aircraft's interior: What kind of shape, color, furniture, space, etc.? Then the desirable physical conditions were written up. Based on this reference, Professor Nagamachi told Glenn Mazur to ask several designers to portray the room interior in several ways to complete the Kansei analysis and formal study.

On the Boeing side, a Kansei engineering research team was formed including Glenn, as the designer and human engineer, a next-generation aircraft construction engineer, a psychologist, an industrial designer, an engineer, and a statistical technician. They planned to have a 2-day Kansei engineering workshop led by Glenn, who had received advice from Professor Nagamachi.

By using the brainstorming of Kansei deployment for Mazda and Milbon of Osaka's adoption, with "relax" as the top concept, the expansion of tree-like words emerged. This Kansei term is expanded from the lower terminology. Next, to meet the top concept, image views of the windows and ceiling, lighting, furniture, walls, seats, etc., were created. In the discussion by the team as a whole, six types of room interiors were chosen (Figure 5.40).

Sixty testers were chosen by using a local recruitment agency to interview them. Also, using the advice of the in-house technical person, a three-dimensional screen dome was produced, and as stated previously, six images were captured and evaluated by Kansei words. By statistically analyzing the data, it was decided to appropriate room conditions in Kansei words.[6]

FIGURE 5.40
Six types of 7E7 interior design samples.

The final data from Jeanne Guerin, a psychologist in Boeing's Design Department were transferred to the Teague Team as decided by the actual interior design based upon the expertise of Professor Nagamachi and his team members. Starting with the cockpit design, also shown were a reclined seat that did not affect the rear seat, the kitchen design, the color of the ceiling and passageway, the expansion of windows, etc., where a relaxed emotion was utilized everywhere (Figures 5.41 to 5.43).

In addition, the idea of All Nippon Airways dispatch technicians was incorporated in these designs, for example, installation of hot water toilets (however, opening and closing of the bathroom door was a failure). As the room design of Boeing 7E7 (later the 787 Dreamliner) is for Europe, BMW also participated in the design. Windows, for viewing the sky, become large, and when you press the black button at the bottom of the window, the room becomes dark even though there is no curtain. This is due to an energized discoloration gel that has been injected into the window, where you can cut off the light from the outside by pressing the button. The entrance and the inside of the 787 aircraft became the innovative design. An easy-to-maneuver-in cockpit was also an improvement.

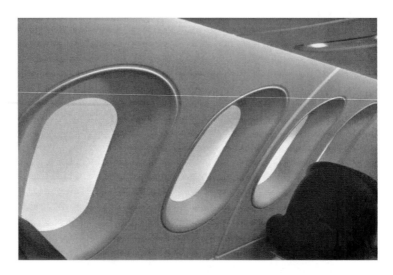

FIGURE 5.41
Boeing 787 and the large type of windows.

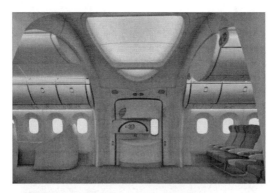

FIGURE 5.42
Entrance and inside of 787.

FIGURE 5.43
The innovative cockpit design.

References

1. M. Nagamachi, I. Senuma, and R. Iwasige. The research of emotion technology. *Human Engineering*, 10(5), 121–130, 1973.
2. M. Nagamachi. Room environment related to emotional analysis. *Human Engineering*, 13(5), 7114, 1977.
3. M. Nagamachi. Kansei engineering related to the sound of the words. *Journal of the Acoustical Society of Japan*, 49(9), 638–644, 1993.
4. M. Nagamachi and M. Yukihiro. Study on sound evaluation of Kansei words using a hierarchical fuzzy integral. Heisei 7 Science and Technology Grant-in-Aid for General Research (B) (Representative Hosoda Kazumasa), 1996.
5. M. Nagamachi. *Story of Kansei engineering.* Tokyo: Japanese Standards Association, 1995.
6. M. Nagamachi (Ed.), *Kansei/affective engineering.* CRC Press, Boca Raton, FL, 2010.
7. M. Nagamachi et al. Development of a pressure-ulcer-preventing mattress based on ergonomics and Kansei engineering. *International Journal of Gerontechnology*, 11(4), 513–520, 2013.
8. J. Guerin. Kansei engineering for commercial airplane interior architecture. Presented at the 16th Symposium on Quality Function Deployment, 19–26, 2004.

6

Global Strategy of Kansei Engineering

6.1 Interaction with South Korea

In 1985, Professor Li Shun from the Faculty of Engineering of Korean University in South Korea intended to further his Ph.D. study and visited the Nagamachi laboratory of Hiroshima University. He had quite a hard time until 1990, when he learned of Professor Nagamachi's expertise at Hiroshima University. To earn his Ph.D. in Kansei engineering was one of Li's fields of study, and after that he published *Kansei Engineering of Information the Age* in Korean as a co-author with Professor Nagamachi and maintained the close relationship. In Korea, Professor Li is one of the many passionate researchers studying Kansei engineering, and Professor Nagamachi frequently visits the society, universities, and companies in Korea and often delivers presentations.

Through the influence of Professor Li, the South Korean government is also interested in Kansei engineering and aimed to promote industries in Korea with Kansei engineering by planning a big project, as shown in Figure 6.1, that is, the Kansei engineering technology development, and provided the activities a budget of 100 billion won (100 million yen). Forty-seven research organizations participated, including five conglomerates of Hyundai, Samsung, LG, etc., and the University of Korea, Seoul National University, and Yonsei University, where Professor Nagamachi became an advisor for this project. Professor Nagamachi has also developed a small car and a dishwasher with LG. This activity continued for 10 years, and Korean companies now have a good understanding of how to make a new product development center based on customer needs.

Professor Nagamachi has claimed that the current Japanese electric product companies have already been lost entirely to Korean companies, and its automobiles industries are also catching up with Toyota. In 1996, Professors Li and Nagamachi together held the Japan-Korea International Society of Kansei Engineering, an international conference in Japan, and in 1997 also held an international conference in Korea, which brought many researchers of Kansei engineering and established the Korea Kansei Engineering Society in 1997. With a delay of 1 year, the Japan Society of

FIGURE 6.1
A professional brochure of the Korean government (1992 to 2001).

Kansei Engineering was born in 1998 in Japan. However, in the following years, since the word *Kansei* originated from the Japanese language, the Korean government declared that it could not be used in Korea, and later they used research activities under a different name. The word *Kansei* originally came from a Chinese word, and it is a word that was brought to Japan through the Korean Peninsula, which Professor Nagamachi greatly regrets.

6.2 Interaction with Europe

In 1996, Professor Nagamachi's close Swedish friend, Professor Jorgen Eklund from Linkoping University, who teaches ergonomics, together with six students, visited Kure National Institute of Technology, where Nagamachi was the president, and conducted a 3-day Kansei engineering seminar. The relationship continues until today, and they conduct lectures a few times at the university, coordinated international conferences, and the students exchanged research presentations. When the 9/11 New York terrorist attack occurred, the international conference of Kansei engineering continued. Although the Europe Kansei Engineering Conference was held twice in Sweden and gathered many young researchers, affecting a lot of Japanese technologies, lack of understanding of the organizer caused the conference to be discontinued. Professor Eklund moved to the Royal Institute of Technology in Stockholm recently and wanted to contribute to *Kansei/ Affective Engineering* (CRC Press, 2010). At a forklift manufacturer (affiliated with the Toyota weaving machine) by BT of Sweden, they developed a forklift that can drive comfortably (Figure 6.2).

The Linkoping University, the Faculty of Engineering in the University of Oulu in Finland, and the University of Leeds in the UK have diligently incorporated Kansei engineering for graduate students working on their Ph.D. theses. Professor Tom Child from the Department of Mechanical Engineering at the University of Leeds possesses a world mastery of mechanical engineering and is attracted to Kansei engineering and focuses on the extension.

Spain is more active. I2BC (Innovation Institute of Human Life and Health) was established in Malaga, Andalucía, and Professor Nagamachi has often been invited by the Institute to establish the Kansei Engineering Research Center. Professor Nagamachi visited there several times and was responsible for training the research group, as well as teaching at the University of Granada and Malaga University. Based on the plan, Professor Nagamachi and the institute gathered 500 industrial companies in the 12BC industrial park. Although the I2BC Kansei Engineering Research Center had previously

FIGURE 6.2
BTicino operation of BT's.

been thought to play the main role in the new product development, due to Spanish government financial problems, the budget was suspended, and currently the building construction has stopped and the activity cannot be executed.

At the University of Oviedo in North Spain and the opposite of Malaga, in the city of Gijon Asturias, Kansei engineering also has been incorporated into the design actively. A group from human engineering offices in Gijon always asks for visits to guide them in Kansei engineering.

When you turn your eyes to Asia, the next active country after South Korea is Malaysia. Professor Nagamachi's apprentice, Associate Professor Dr. Anitawati Mohd Lokman, led the Kansei engineering ventures at the University of Technology, MARA. When she planned to get her Ph.D., Professor Nagamachi coached for more information on and off site, and she successfully completed her Ph.D. Since then, many of the graduate students have learned Kansei engineering and would like to study in Japan, but there is no university to accept them (they need scholarships), and the universities are also in trouble. The government is also keen to introduce Japanese technology through the Look-East Policy, but financially it is having problems.

In America, despite receiving huge applause when Mazda president Kenichi Yamamoto gave a special lecture on Kansei engineering at the University of Michigan in 1986, there has been no response. Professor Nagamachi has promoted Kansei engineering and has often been invited by General Motors and Ford, but the only output that we can see is the Boeing 787. On the other hand, in Mexico, Professor Nagamachi's apprentice Ricardo Hirata is working hard, giving lectures to two universities, Autonomous University of Mexico (300,000 students) and Ibero Americana University (100,000 students). And in Techno Park in Toluca City, where the world of the automobile company is focused, there have been talks of the establishment of a Kansei engineering research center, but this has not been realized yet due to insufficient funding.

6.3 Development in Asia

Back to Asia, Professor Nagamachi presented two Kansei engineering lectures at Xi'an Jiaotong University's Faculty of Engineering in China and received an incredible response from graduate students and faculty. However, Taiwan has become a hot spot in Kansei engineering. The third sector of the Taiwanese government Ministry of Economic Affairs has frequently invited Professor Nagamachi to the Zhongwei development center (staff of 360) to guide them on how to use the industrial technology of Japan in Taiwanese companies, and to conduct lectures at the university

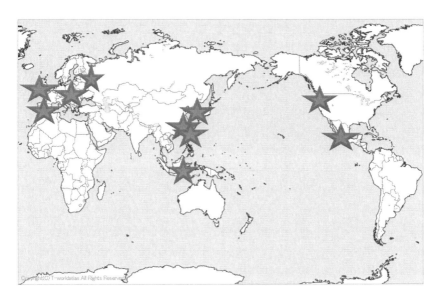

FIGURE 6.3
Global spread of Kansei engineering.

and company. The manufacturing industries cover bicycles, cars, car spare parts, tile, etc., and the universities include Zhongyuan University, National Taichung University of Education, and MotoSatoshi University. The Taiwan Industrial Technology Research Institute (ITRI) (staff of 6000) established the Kansei Engineering Research Division. It is possible for Kansei engineering to develop rapidly in Taiwan.

The countries that Professor Nagamachi has visited are marked with an asterisk in Figure 6.3. It illustrates that Kansei engineering has spread throughout the world.

7

Growing Kansei

Below is the methodology of Kansei engineering:

1. Behavior observation: Developed Sharp refrigerator, LCD ViewCam.
2. Kansei engineering type 1: Developed Wacoal bra, Milbon shampoo, and pressure ulcer prevention mattresses.
3. Kansei engineering type 2 (sensibility ergonomics): Developed toilet of Matsushita Electric Works (TRES).
4. Kansei engineering type 3 (artificial intelligence system): Developed virtual house design system kitchen of Matsushita Electric Works, virtual house design with Kansei Electric Power Company, Komatsu cockpit, interior design of Isuzu Motors, Ltd., brand name diagnostic system, Kansai Electric Power.
5. Kansei engineering type 4 (rough set theory model): Children's shoes.

You will be a Kansei engineer if you read the general references found at the end of this chapter. Kansei needs to be polished on our own. If you are not serious about the use of Kansei, it is impossible to develop new products that are welcoming to the market even if you have learned the Kansei engineering approach thoroughly. How to polish your own Kansei is as follows:

1. **Get used to the ergonomic scale.** If you can understand good and bad work environments, you will know whether people work with good feelings or not. Therefore, even without bringing the measurement tool, you can determine by ear "the noise here is XX decibels," and the guidance of environmental improvement is possible. Similarly, when related to lighting, "the brightness here is XX lux," you can improve the lighting, but it is possible from the decision of your own eyes. If you train several times using the instrument, you can scale the brightness and loudness in your mind. Professor Nagamachi always carries a watch with a second hand to measure time.

   Professor Nagamachi always measures the operation activities by using a watch when entering the site, and decides whether the speed of work is appropriate or not. By experiencing the developing pressure ulcer–prevention mattress, if you put any mattress beside him, he will accurately judge whether it is suitable for a good sleep.

The body will be the scale. It is necessary to make an effort to embed the scale of several different types, which allows you to measure the environment using your own body.

If you study a little color adjustment, you will be able to determine the appropriate colors of the buildings of a city, and the same goes for when you enter a supermarket: you will be concerned about the design, color, and shape of the products.

2. **Discard the feeling of likes and dislikes.** When the preference feelings of likes and dislikes are strong, this will damage Kansei judgment. If you like blue because blue is your favorite color, and you hate the shape and that is why you say no—such judgments are looking at the things through one's feelings (Kansei), and you will not notice that your feelings are different from other persons' feelings. It is basically looking at the scale of likes and dislikes. When you reduce the feelings of likes and dislikes, you try to be in a neutral position as an effort to acquire power for the decision.

3. **Have a wide interest.** It is important to have numerous Kansei measurement scales rather than a single measurement scale. Although Mr. X has knowledge experts, one might not be good in another field, and it is inadequate to judge the Kansei of others. People have interests in various things. So in order to assume the Kansei of many people, there is a need to have a variety of interests from different people.

   It is important to excel in culture, nature, and history. You should get used to, to some extent, bright people for sports as well as have an interest in the history of buildings. For people who like designs, particularly costumes for adults, children, and the elderly, it is a good idea to have an interest in personal belongings. It is also hoped that one has a broad knowledge in color, form, and design. It is advised to have various types of scales rather than a single scale, and have a variety of interests and hobbies.

4. **Have a long, precise scale of Kansei measurement.** It is unnecessary to have two scales of a design for measuring the Kansei of good or not good. Multidimensional scales are only required to measure how much good or bad something is.

   When Professor Nagamachi travels with students and points to buildings, he asks them what the Kansei word of that building is. The students will answer, "Luxury." Professor Nagamachi immediately questions: "Luxury, in what point?" Students answer, "8 points in the 10-point scale."

   And then Professor Nagamachi asks the following: "What is your baseline of luxury of 8 points?" The students ponder and answer,

"There are several pieces of roof." Next, Professor Nagamachi points to the ladies walking on the street and asks, "How attractive is that girl if you give points?" The students answer, "6 points for attractiveness." Professor Nagamachi asks again: "What is your baseline of attractiveness?" They answer, "The eyes are round and like chestnuts." Professor Nagamachi asks again: "What is the bad thing?" Then they answer: "The thickness of her legs." Professor Nagamachi always trains his students using this method.

These kind of adjectives, such as *luxury* or *attractiveness*, are called *Kansei words*. Professor Nagamachi encourages you to think of the appropriate Kansei words based on the Kansei evaluation. On the other hand, the number of roofs and the thickness of legs are called the *physical elements*. In summary, the baseline that produced the Kansei is the physical elements. We are trained to be able to accurately determine the points with the appropriate sensitivity word and the physical element as a basis to look at things. A reader also walking down the street and doing a monologue on the thing that flows to his or her eyes, when self-disciplined, will keep his or her focus on the Kansei decision. If possible, find your friends and sometimes several people, and argue back and forth like this, and the exact Kansei will grow.

5. **Sometimes eat those delicious and luxurious foods.** When you enter a restaurant during lunchtime, you see the employees ordering cheap and affordable food. Additionally, you see people buying boxes of lunch side by side in a stall. This is normal for this case, but sometimes it is suggested that you try the French food, Italian food, Spanish food, or Japanese cuisine. The surrounding interior of the eating venue, the excitement of the moment you see the food come out, the taste of when it is placed in the mouth, the food bite by bite, and things such as plating and the stunning color that is served on a plate are a lot of inspiring events.

6. **Go to the gallery or museum.** View an excellent picture that is a scene most suitable to cultivate sensitivity. There are many excellent museums of art everywhere. If you travel overseas, adopt the popular habit of going out to museums or galleries. The Museum of Modern Art and the Metropolitan Museum of Art in New York, the Washington National Gallery, the Art Institute of Chicago, the British Museum of London, the Louvre in Paris, the Picasso Museum and Dali Museum in Spain, and the Design Museum of Copenhagen are various excellent places to go. Such design, art, crafts, etc., develops you to have a rich sensibility.

The Tokyo Motor Show or Detroit Motor Show are collaborations of many people from each company, not the work of one person.

The ability to absorb the atmosphere, the sensibility of design forces, and concepts of each car manufacturer are important. Also being exposed to a situation like this will give you a chance to cultivate excellent sensitivity.

General References for Kansei Engineering

1. M. Nagamachi. *Kansei engineering*. Kaibundo Publication, Tokyo, 1989.
2. M. Nagamachi (Ed.). *Kansei Product Science: Fundamentals and Applications*. Kaibundo Publication, Tokyo, 1993.
3. M. Nagamachi. *Compilation. Kansei commodity science*. Kaibundo Publication, Tokyo, 1993.
4. M. Nagamachi. *Story of Kansei engineering*. Japan Standards Association, Tokyo, 1995.
5. M. Nagamachi. *Kansei and product development*. Kaibundo Publication, Tokyo, 2005.
6. A. Lokman and M. Nagamachi. *Innovation of Kansei engineering*. CRC Press, Boca Raton, FL, 2010.
7. M. Nagamachi (Ed.). *Kansei/affective engineering*. CRC Press, Boca Raton, FL, 2010.

Index

A

Abeyama (Mr.), 24, 44
After Eight chocolate, 82–87
aging
    bedsores, 88
    hospitality, 38–39
    population ratios, 64
    toilet deign, 74–75
Aging of Society Policy, 9
air conditioners, 26
airplane interior design, 95–99
All Nippon Airways, 97
America, 104
amoeba management, 41–43
Aoki managing director, 61
articulation, 75–76
artificial intelligence, 6, 79–81, 107
Art Institute of Chicago, 109
art museums, 109
"Asahi" newspaper, 54
Asia, global strategy, 104–105
auditory stimulation, 2
automated transportation system, 39
Avanse Series 1 (PC45), 63

B

bath towels example, 50
beautiful or well-balanced feeling, 2
beautiful woman example, 1
bedsores, 88–95, 107
Beethoven, L. van, 2
behavior observation, 6, 107
bending at waist, *see* Posture
big data, 6
blood flow, 92–93
BMW company, 97
body pressure dispersion, 89–92,
    see also Bedsores
body pressure variance, 74
Boeing 787 interior design, 95–99
brand name decision, 75–82, 107
Breathair material, 89–95

breathing method, 75
British Museum of London, 109
BT (Sweden), 103
bullying, 5
Bungei Shunju, 76

C

Canon company, 26, 28
cars
    one-man assembly work, 20, 22
    ride quality example, 1, 2
cell production system, *see also*
        Employees
    ergonomic improvements, 31
    manufacturing application, 18–27
    Mitsubishi Electric company, 45
cerebrophysiological method, 6
Chaplin, Charlie, 15
Chevron, 46
Child, Tom, 103
children's shoes, 107
child with puppy example, 1
Chrysler company, 15
classical music example, 2
"clean heater" method, 24
coal mining example, 16
cockpit design
    Boeing 787, 97, 99
    Komatsu, 61–64, 81, 107
coins example, 45
"collection of rivals," 44
color coordination system, 81
communication, 46–48
composite sensation, 2
concealed emotions, 5
confectionery package design
    conducting evaluation research, 84
    gathering research candidates, 83
    Kansei words, 84
    overview, 82
    package design presentation, 86–87
    survey data analysis, 85